# In Search of "Aryan Blood"

# CEU Press Studies in the History of Medicine

## Volume IV

Series Editor: Marius Turda

# In Search of "Aryan Blood"

*Serology in Interwar and National Socialist Germany*

**Rachel E. Boaz**

C E U PRESS

Central European University Press

Budapest—New York

© 2012 by Rachel E. Boaz

Published in 2012 by
Central European University Press
An imprint of the Central European University Limited Company

Nádor utca 11, H-1051 Budapest, Hungary
*Tel:* +36-1-327-3138 or 327-3000 · *Fax:* +36-1-327-3183
*E-mail:* ceupress@ceu.hu
*Website:* www.ceupress.com

400 West 59th Street, New York NY 10019, USA
*Tel:* +1-212-547-6932 · *Fax:* +1-646-557-2416
*E-mail:* mgreenwald@sorosny.org

ISBN 978 9639776500

ISSN 2079-1119

Library of Congress Cataloging-in-Publication Data

Cassata, Boaz, Rachel E.
In search of "Aryan blood" : serology in interwar and National Socialist Germany / Rachel E. Boaz.
p. cm. (CEU Press studies in the history of medicine, ISSN 2079 1119 ; v. 4)
Includes bibliographical references.
ISBN 978 9639776500 (hardbound)
1. National socialism and medicine History. 2. Serology Political aspects Germany History 20th century. 3. Anthropology
Political aspects Germany History 20th century. 4. Racism in medicine Germany History 20th century. 5. Racism in
anthropology Germany History 20th century. 6. Biopolitics Germany History 20th century. 7. Antisemitism Germany History
20th century. 8. Germany Politics and government 1918 1933. 9. Germany Politics and government 1933 1945. 10. Germany
Ethnic relations History 20th century. I. Title.
R510.B63 2011
616.07950943 dc23
2011037678

Printed in Hungary by Akadémiai Nyomda, Martonvásár

# CONTENTS

# LIST OF FIGURES

# ACKNOWLEDGEMENTS

I received support and assistance from many in writing this manuscript. For their guidance and advice, I want to thank Richard A. Steigmann-Gall, Shelley O. Baranowski, Elizabeth M. Smith, and Alison Fletcher. Marius Turda and Paul Weindling provided both invaluable judgment and sharing of information. I am indebted to both Marius and Nóra Vörös for their patience and careful editing. There are numerous librarians and archivists to whom I am grateful for their assistance: Anja Adelt (Hauptstaatsarchiv Stuttgart), Andreas Grunwald (Hauptstaatsarchiv, Berlin), and Jonathan Hartnett, Vincent Slatt, and Ron Coleman (all at the United States Holocaust and Memorial Museum). I am further indebted to Kent State University's interlibrary loan department for handling dozens of my requests. Of the authors whose work I have cited throughout the text, William H. Schneider, Michael G. Kenny, Douglas A. Starr, and Pauline M. H. Mazumdar were especially helpful. For their advice and encouragement, I am further indebted to my colleagues and friends Monika Flaschka and Erika Briesacher, and my family and Zachary for their moral support.

# INTRODUCTION

In Nazi Germany, basic civil rights—and ultimately the right to live—depended on whether one had "German blood." Meticulous racial categorization of individuals as either "German-blooded" or "non-German-blooded" relied primarily upon documentation, such as birth and baptismal certificates. Labeling was compulsory, as everyone was required to carry identification indicating their race. However, in cases where parentage was in dispute or the necessary records were missing, the state often referred the matter to so-called "racial experts." In one such instance in 1940, Fritz Lenz explained to a district court that he could not give an expert opinion without first testing the subject's blood type. The Ministry of Justice became involved and referred the matter to several of Lenz's colleagues, who replied that they did not share his views at all and that they had no difficulty in making an accurate assessment on the basis of photographs alone.[1] This was not uncommon, as physiognomic, or visible, characteristics were the most common means of differentiation between disparate "racial types." As examiners' preferences varied, determining race was often a highly subjective process. Some "experts" maintained that race was apparent in the shape of one's skull, nose, or jawline, while still others looked at the pigmentation of the hair, skin, and eyes. In the end, it was most often a combination of these that went into a final decision. In spite of the critical "sorting" of individuals based on blood, the trait of blood type was very rarely referenced in an anthropological context.

---

[1]  Benno Müller-Hill, *Murderous Science: Elimination by Scientific Selection of Jews, Gypsies, and Others, Germany, 1933–1945* (New York: Oxford University Press, 1988), 40.

In the decade just before Hitler came to power, a branch of medicine developed that sought to identify race through blood. Seroanthropology, as it was called, was a meeting of two sciences: serology—the study of blood—and anthropology. This interesting blend led some researchers to affiliate blood type with a host of other characteristics, some of which seemed related to race and others not. In 1930 one examiner claimed that persons with type O blood had the best teeth, followed by those with type AB, whereas type A and B individuals were supposed to have the worst.[2] Another declared that type O subjects had "less satisfactory" strength of character and personality, and that group B individuals were impulsive.[3] Other associations were made with career choice, sadism, and even flat feet. Later came still further obscure affiliations between blood type, IQ, and even the feeding preferences of mosquitoes.[4]

During the interwar period, contributions from a range of disciplines created a diverse group of racial theorists. Different perspectives gradually created the extremes of "*völkisch*" and "non-*völkisch*" philosophies. *Völkisch* theory was associated with what were referred to as "negative" eugenic measures, such as sterilization or clinical confinement. This proposed restriction of civil liberties was often paired with a belief in the inherent superiority of certain races over others. By contrast, "positive eugenics" concentrated instead on social and medical assistance.[5] Seroanthropology, with its implicit emphasis on race, was particularly well received by German researchers interested in applying this new science in order to substantiate their own racial biases. The prospect of a simple blood test for racial identification would have been attractive. By World War I there was widespread disillusion and frustration with skull measurements, certainly the most common racial diagnostic, and it was hoped that blood might provide a more efficient methodology.[6] These expectations led to an impressive amount of seroanthropological research globally, but especially in Germany. This book traces the course of German seroanthro-

---

[2]  George Garratty, *Immunobiology of Transfusion Medicine* (New York: Marcel Dekker, 1993), 201.
[3]  Ibid.
[4]  Ibid.
[5]  Marius Turda and Paul J. Weindling, "Eugenics, Race and Nation in Central and Southeast Europe, 1900–1940: A Historiographic Overview," in Marius Turda and Paul J. Weindling, eds., *Blood and Homeland: Eugenics and Racial Nationalism in Central and Southeast Europe, 1900–1940* (Budapest: Central European University Press, 2007), 11.
[6]  Christopher M. Hutton, *Race and the Third Reich* (Malden, MA: Polity Press, 2005), 23.

pology from its promising beginnings after World War I to its neglect during the Third Reich.

The dismissal of seroanthropology, a branch of racial science, is all the more interesting when one considers how important the concept of "blood" was figuratively, and then literally, to National Socialism. From the beginning, the Nazis had consistently exploited a notion of scientific "blood difference" in their propaganda. Use of the term blood (*Blut*) was politically productive; one could evoke ties of blood and imply racial kinship without directly invoking the problematic concept of anthropological race.[7] Nazis commonly referred to *Bluteinheit* ("unity of blood"), *Blut und Boden* ("blood and soil"), *Blutbewusstein* ("blood or racial consciousness"), *Blutsgemeinschaft* ("community bound together by blood ties"), and so on.[8] In doing so the Nazis were simply taking advantage of the mystical significance of blood common to so many cultures and peoples. The Egyptians saw blood as the carrier of the vital human spirit and would bathe in the liquid as a restorative.[9] The Roman historian Pliny wrote about the custom of epileptics drinking the blood of wounded gladiators, treating the latter as if they were "live cups."[10] Blood could be life-giving but also destructive. Like many tribal peoples, the Cherokees believed that the debilitating power of menstrual blood could be channeled against an enemy and thus was often evoked in war, sorcery, and ball game rituals.[11] The Bible mentions blood more than 400 times, and this is reflected in the representative importance of blood in both Judaism and Christianity to this day.[12] It was easy for the Nazis to play upon the symbolism of blood, but the use of such symbolism did not stop there. It eventually affected both their social and medical policies.

A veritable avalanche of scholarship has explored in detail the racial, welfare, and medical policies of the National Socialist regime. As one author has described it, the scholarship has moved Nazi racial and eugenic thought

---

[7]   Ibid., 18.

[8]   Ibid.

[9]   Mino Gabriele, "Magia Sanguinis: Blood and Magic in Classical Antiquity," in James M. Bradburne, ed., *Blood: Art, Power, Politics, and Pathology* (New York: Prestel, 2002), 36.

[10]   Ibid.

[11]   Circe Sturm, *Blood Politics: Race, Culture, and Identity in the Cherokee Nation of Oklahoma* (Berkeley: University of California Press, 2002), 34.

[12]   Douglas A. Starr, *Blood: An Epic History of Medicine and Commerce* (New York: Knopf, 1998), xiv.

and policy to the center of our understanding of the nature and dynamic of the Third Reich.[13] Curiously, though many aspects of Nordic race theory have been scrutinized, racial studies of blood have not. Numerous studies refer to the Nazis' tendency to conflate ideas of blood and race, but focus on the metaphor of blood instead of its application to medical research.[14] There have been only articles—no monographs—addressing German seroanthropology. Pauline M.H. Mazumdar's "Blood and Soil: The Serology of the Aryan Racial State" touches on the implications of seroanthropological research to a specific group of race theorists. However, she deals primarily with the mechanisms of racial studies of blood and provides only a cursory analysis of its theoretical implications.[15] Myriam Spörri demonstrates how the popular idea of "Jewish blood" as somehow different and contagious was manifested in German medicine prior to the Third Reich.[16] She discusses briefly how seroanthropology, like other branches of race science, was characterized by political extremes. I, too, address this division, but focus mainly on the appropriation of blood science by the radical right. Seroanthropology merits study not simply because it is an unexplored discipline within racial anthropology, which is intriguing in and of itself, or because such an analysis would fill a gap in the research, but because the fate of seroanthropology, namely its dismissal by the Nazis, contradicts the current "biopolitics" argument within the broader historiography of German medicine. This theory draws a line of continuity between pre-1933 developments in Germany and the Nazi atrocities eventually committed under medical pretenses. Historians Detlev Peukert, Zygmunt Bauman, Peter Fritsche, and Thomas Rohkrämer have all focused on the relationship between modern scientific biopolitics and National Socialist racial policy.[17]

---

[13] Edward Ross Dickinson, "Biopolitics, Fascism, Democracy: Some Reflections on Our Discourse About Modernity," *Central European History* 37 (2004): 4.

[14] Allyson D. Polsky, "Blood, Race, and National Identity: Scientific and Popular Discourses," *Journal of Medical Humanities* 23, nos. 3–4 (December 2002): 171–186. Polsky states that "National Socialist physicians used blood (and other traits) to determine the racial identities of those with 'impure' origins" (ibid., 172). See also Uli Linke, *Blood and Nation: The European Aesthetics of Race* (Philadelphia: University of Pennsylvania Press, 1999).

[15] Pauline M.H. Mazumdar, "Blood and Soil: The Serology of the Aryan Racial State," *Bulletin of the History of Medicine* 64 (1990): 187–219.

[16] Myriam Spörri, "'Reines Blut,' 'gemischtes Blut': Blutgruppen und 'Rassen' zwischen 1900 und 1933," in Anja Lauper, ed., *Transfusionen: Blutbilder und Biopolitik in der Neuzeit* (Zurich: Diaphanes, 2005), 211–226; and Myriam Spörri, "'Jüdisches Blut': Zirkulationen zwischen Literatur, Medizin und politischer Presse, 1918–1933," *Österreichische Zeitung für Geschichtswissenschaften* 3 (2005): 37–38.

[17] Dickinson, "Biopolitics, Fascism, Democracy," 6.

Peter Fritzsche has referred to the "dark shadows of modernity" lurking in the Weimar Republic.[18] Peukert has similarly claimed that the "scientization and medicalization" of social problems in the pre-1933 period had murderous results during the Third Reich.[19] He has correctly outlined how anthropological racism, with anti-Semitism as its centerpiece, became radicalized in the following stages: bans on emigration; deportations to the East; unsystematic mass killings; and, finally, systematic mass killings.[20] Edward Ross Dickinson believes that the historiography has been, collectively, so focused on unmasking the "negative potentials and realities" of this modernity that we have constructed a "true, but very one-sided picture."[21]

This book proposes a different historical vision. I examine seroanthropology as a divergence from this model of biopolitics. It does not fit neatly into this paradigm and thereby urges a rethinking of the "dark potential" of pre-Nazi race science. Gross misappropriation of racial anthropology is, in part, to blame for the genocide that occurred. Like conventional race science, blood science was also misused, and attempts were made to apply it with the same malevolence. Yet no one has taken much notice of this, despite the fact that seroanthropology involved a large-scale research effort; hundreds of studies were conducted, and tens of thousands of German subjects had their blood types tested for racial purposes between 1919 and 1939.[22] Even so, racial analyses of blood were exceptionally rare by the end of the interwar period, and seroanthropology was rejected by the mainstream German medical establishment even years before that. Why it was abandoned by the Nazis is one of the underlying questions of this work.

This book is organized both chronologically and thematically. Chapters Two through Five follow the development of seroanthropology until 1933. Chapter Two examines developments in the early twentieth century

---

[18]  Peter Fritzsche, "Did Weimar Fail?" *Journal of Modern History* 68 (1996): 632 and 649.

[19]  Detlev Peukert, "The Genesis of the 'Final Solution' From the Spirit of Science," in David Crew, ed., *Nazism and German Society, 1933–1945* (London: Routledge, 1994), 275.

[20]  Ibid., 277.

[21]  Dickinson, "Biopolitics, Fascism, Democracy," 25.

[22]  "Between 1919 and 1939, the total number of international subjects whose blood was typed and reported in published articles was 1,354,806. A systematic search of the major medical and anthropological journals published in the United States, Britain, France, and Germany indicates that this comprised over 1,200 articles." William H. Schneider, "The History of Research on Blood Group Genetics: Initial Discovery and Diffusion," *History and Philosophy of the Life Sciences* 18, no. 3 (1996): 287 and 280.

that led to the modern discipline of serology. Slow and erratic progress in blood science was paralleled by a renewed interest in studies of race. In an attempt to emphasize "biological" differences between racial types, anti-Semitic literature of the time borrowed from both literature and science. This hazy blend of blood superstitions and clinical objectivity would persist into Nazism. I further examine the social context in which this new science was formed to analyze why physicians of Jewish lineage played such a critical role in the formation of not only serology, but seroanthropology as well—the main purpose of which was to identify a relationship between blood and race.[23] John Efron has emphasized the "intellectual resistance" of Jewish racial anthropologists in the early twentieth century.[24] However, because of the intellectual climate at this time and the fact that race science was viewed as a perfectly legitimate pursuit, I show how Jewish involvement in studies of race is more nuanced than Efron has claimed. Chapter Three examines the diffusion of seroanthropology in the 1920s. Racial studies of blood were at their height in the decade after the war, when researchers were most eager to apply them to escalating racial and eugenic concerns. I examine the efforts of racially biased practitioners to relate their findings to a political agenda of anti-Semitism, nationalism, and xenophobia. For certain race scientists, politics and medicine were inextricably linked, and the circumstances in postwar Germany helped to make seroanthropology attractive to individuals with this tendency. They were initially hopeful that studies of race and blood might be useful in identifying "racially pure" Germans, excluding "racial others," and even assisting in cleansing the German populace of those considered genetically unfit. Simply put, they wanted to base common references to "differences in blood" on solid biological principles. It was largely due to the political circumstances into which seroanthropology was introduced that Germany was the nation in the "forefront of developments" in the field of blood type studies during the interwar period.[25] Chapter Four details the establishment of the German Institute for Blood Group Research, whose directors, physicians Otto Reche and Paul Steffan, were primarily concerned not so

---

[23]   Robert Proctor, *Racial Hygiene: Medicine under the Nazis* (Cambridge, MA: Harvard University Press, 1988), 1.

[24]   John M. Efron, *Defenders of the Race: Jewish Doctors and Race Science in Fin-de-siècle Europe* (New Haven: Yale University Press, 2004).

[25]   Schneider, "The History of Research on Blood Group Genetics," 295.

much with the developing clinical or legal applications of blood science but with how it might contribute to *their* existing understandings of race. Individuals who fell under the grouping of "seroanthropologists" were a mixed lot. Even though a medical degree was considered the most appropriate path of study for "race scientists," many who studied blood and race were not blood scientists or even racial anthropologists.[26] As a brand new field of study, and one with potential political consequence, seroanthropology had a diverse range of practitioners. Chapter Five examines the institute's priority of outlining the serological makeup of Germany and locating the most "racially pure" Germans. By analyzing blood type, Reche and Steffan hoped to determine the extent to which Eastern, or "non-German," blood had spread into Germany. Their colleagues would examine blood and its possible relation to inferior genetic traits.

The first five chapters trace changes in seroanthropology until the late years of the Weimar Republic. Chapter Six continues from this point but progresses into the early years of the Third Reich—immediately prior to the introduction of the 1935 Nuremberg Race Laws. I discuss the role of Jews in relation to seroanthropology by analyzing racial studies of blood conducted by researchers of Jewish descent, but also by looking at studies in which Jews were the subject group. National Socialist propaganda made frequent use of "blood rhetoric" and, like the anti-Semitic literature of the pre-1918 period, distorted blood science to confirm supposed racial difference. I further examine why Jews, even in the era of National Socialism, continued to contribute to a discipline intent on distinguishing between racial types. Chapter Seven deconstructs the Nuremberg Laws and how blood was used in their abstractly defined categories. This legislation demonstrates the Nazis' belief that "blood defilement," through both sexual and clinical means, was a serious threat. The Nazis made persistent references to "differences in blood" between races and legally acted upon these assertions, even when they knew the assertions were false. This remains an unaddressed hypocrisy of the Third Reich. Imprecise categories of "blood difference" were used not only in the construction of racial classifications but also in their interpretation. Chapter Eight continues this discussion of disparities in blood but examines how these changed under the circum-

---

[26] Proctor, *Racial Hygiene*, 8.

stances of war; the discourse used in the Nuremberg Laws was now applied to the Eastern occupied territories and the Nazis' intent to reclaim "German blood" for Germany. The war, however, also affected the priorities of German blood science. Frustrated by conventional studies of blood and race, physicians pursued different avenues of seroanthropological research made possible by German conquests. At the same time, the concerns of the military and the courts dictated more practical uses of blood, though their efforts were often too little, too late when compared to the methodologies of the Allied powers.

Seroanthropology is marked by interesting contrasts. It was introduced by Jewish scientists, but quickly came to be used for anti-Semitic purposes; blood science was commonly referenced in propaganda, but this usually required embellishment or outright lying; it drew both political and apolitical practitioners; and Germany expressed interest in it that was nowhere near as pronounced in other "medical giants," such as Great Britain and the United States. These continuities and discontinuities are apparent throughout this book and serve to broaden our understanding of race science in the early twentieth century.

# CHAPTER I

# THE EMERGENCE OF BLOOD SCIENCE

Since its primitive beginnings in early modern Europe, racial anthropology traditionally relied upon physiognomic traits to determine an individual's race. Physiognomic traits meant bodily characteristics, which often included measurements of one's nose, various dimensions of the skull, and/ or the pigmentation of the hair, skin, and eyes.[27] The notion that race was evident in an individual's appearance was accepted among racial anthropologists in general, but it was especially promoted by *völkisch* race theorists from their origins in the pan-German movements of late-nineteenth-century Europe until the end of the Third Reich. The later years of Imperial Germany and the early Weimar Republic, which were accompanied by an intensification of right-wing race propaganda, underlined the idea that race manifested itself in appearance. This tendency was the result of a gradual "mathematization of medicine and anthropology" in which comparative craniometry, the measuring of skulls, figured most prominently.[28]

The early National Socialist movement, as well as other racist groups that emerged from the discontent of World War I, often borrowed from existing notions of economic, political, cultural, and racial anti-Semitism. However, race proved especially useful as a political tool. Extremist propaganda distorted medical facts in order to claim racial difference and justify

---

[27] Pieter Camper (1722–1789) presented the "facial angle" as a method of racial comparison. Europeans were at the top of a scale which included both animals and humans—the larger the angle, the "higher" the race. For example, a European might have a measurement of 80 degrees, and an African 70. Certain types of primates could reach 50 degrees. Therefore, a lower facial angle suggested an individual was, in fact, less evolved (and racially inferior). Shortly after Camper, Anders Retzius (1796–1860) developed the concept of the "cephalic index." Proctor, *Racial Hygiene*, 12.

[28] Efron, *Defenders of the Race*, 15.

segregation and curtailing of the rights of those viewed as "racial others." It also made frequent mention of another, less obvious racial indicator—that of blood. Some of the early stereotypes of racial physiognomy were gradually supplemented by the relatively elusive "racial characteristic" of blood. Several works from this time touch on how propagandistic notions of blood, similar to physiognomic differences, were portrayed as biological fact. Although they were written not as anti-Semitic propaganda, but rather as a commentary thereof, the stories by Oskar Panizza and Salomo Friedlander are useful because they depict the growing social problem of scientific racism in such a misrepresented and offensive manner—similar to Artur Dinter's anti-Semitic fiction work *Die Sünde wider das Blut* (The sin against the blood). References to clinical procedures in blood science began to appear in racial propaganda at this point because of actual developments within medicine.

Ironically, political groups were using medical notions for anti-Semitic purposes at a time when Jews were highly visible members of Germany's medical community. This was especially the case in serology; physicians of Jewish descent established the discipline, as well as its later application to racial studies. In spite of the "more pervasive and organized anti-Semitism in Germany and Austria" compared to Europe as a whole, both nations had a healthy representation of educated and prominent Jews. [29] Karl Landsteiner, an Austrian Jew who later converted to Roman Catholicism, made modern serology possible through his discovery of the blood types in 1900. The racial analysis of blood, however, was not practiced until after World War I. For racial theorists, wartime population shifts provided an advantageous set of circumstances in which different peoples could be studied, thus allowing Polish physicians Ludwik and Hanna Hirszfeld to conduct extensive blood type surveys of various "racial types." The results of their work implicated blood's anthropological, or racial, significance.

"Contagious Blood" in German Fiction and Early Blood Science

Published in 1893 by physician Oskar Panizza, the fictional story *Der Operierte Jud* (The operated Jew) details the extensive efforts of a German

---

[29]  Ibid., 10.

Jew, Itzig Faitel Stern, to lose his Jewish identity. It begins with a grotesque physical description of Stern's Jewishness:

> Itzig Faitel's countenance was most interesting. It is a shame that Lavater had not laid eyes upon it.[30] His lips were fleshy and overly creased; his teeth sparkled like pure crystal. A violet fatty tongue often thrust itself between them at the wrong time. If I may also add that my friend's lower torso had bowlegs whose angular swing was not excessive, then I believe that I have sketched Itzig's figure to a certain degree.

The author continues, detailing a lengthy series of procedures to correct what are the supposed physical manifestations of "racial inferiority." The operations, which include "new legs," various surgeries, speech lessons, and hair straightening and lightening, are believed to successfully convert Stern physically into a German. Stern's appearance and mannerisms no longer betray the fact that he is a Jew. However, the procedures do not address the matter of his "racial temperament." Even when fitted with an "Aryan" appearance, Stern still lacks the essential "chaste, undefined Germanic soul."[31] As explained by Panizza, this "*undefined* German soul" (italics mine) was believed to be contained in the blood, since it was "possible to assert to a certain degree that the abode of the soul could be located in the blood and its changing condition."[32] To remedy this absence, Stern proposes a transfusion of "racially German" blood. Though advised not to go through with a dangerous procedure that has "fallen out of fashion," Stern remains adamant in his intent to "buy [me] some Chreesten blud!" and proceeds to pay six "hardy [German] people" for a liter of blood each. Upon hearing that their blood is destined for a Jewish recipient, however, the German donors abruptly withdraw their offer.[33] The necessary blood is eventually drawn from "seven strong women from the Black Forest." Wanting to shed his Jewishness completely, Stern "let everything [all the blood] run out that could." The eight liters of blood from the German peasant women is then transfused into Stern. Nonetheless, this transfusion does not have

---

[30] Johann Kaspar Lavater (1741–1801), German poet and descriptive physiognomist, published *Physiognomische Fragmente zur Beförderung der Menschenkenntnis und Menschenliebe* (1775–1778), a compilation of essays on physiognomy.

[31] Jack Zipes, trans., *The Operated Jew: Two Tales of Anti-Semitism* (London: Taylor and Francis, 1992), 59.

[32] Ibid., 60.

[33] Ibid., 60–61.

the desired effect; after a few weeks, it is noticed that Stern is "again making new attempts to gain possession of the German soul."[34]

Because propaganda reflected the norms of anthropological racial classification, most racial tracts, including *The Operated Jew*, centered on racial physiognomy. The text diverges from the norm, in its inclusion of the procedure of blood transfusion. In the late nineteenth century, anti-Semitic allusions to a literal relationship between blood and race were becoming increasingly common. A racial blood theory was proposed by French race theorist Arthur Gobineau in his *L'Essai sur l'inégalité des races humaines* (An Essay on the Inequality of the Human Races), published in 1853. A writer and poet with ambitions in the history of philosophy, Gobineau postulated that blood was the decisive factor in the historical and civilizational development of mankind, and that the blood of the Caucasian type—meaning the white race, but particularly the Aryans—was superior to all others.[35] During his racist and anti-liberal campaign, German politician Adolf Stöcker claimed that it was not possible to transfuse "Aryan blood" into the veins of Jews.[36] The Aryan occultist Lanz von Liebenfels claimed that the "non-Aryan" races literally had non-human, or animal, blood.[37] Panizza, too, is guilty of blending clinical practice with unclear notions of race. Stern's literal receipt of "German blood" to remove his Jewishness exemplifies the age-old notion that identity is rooted in the blood. The reality of this metaphor is now implied through an actual exchange of blood.

When *The Operated Jew* was published in 1893, modern blood science was still in its infancy. Though both research and therapy involving blood were practiced, they were not undertaken consistently, and the results were not reliable. Medical treatment concerning blood in medieval and early modern Europe had often involved bleeding, a holdover from the ancient belief that all natural phenomena resulted from the interplay of the four elements—air, fire, water, and earth. The Greeks assumed that four analogous factors must govern the body. These elements, or "humors," included

---

[34]  Ibid., 61.

[35]  Annette Weber, "'Blood Is a Most Particular Fluid': Blood As the Object of Scientific Discovery and Romantic Mystification," in James M. Bradburne, ed., *Blood: Art, Power, Politics, and Pathology* (New York: Prestel, 2002), 164.

[36]  Paul Weindling, *Health, Race and German Politics between National Unification and Nazism, 1870–1945* (Cambridge: Cambridge University Press, 1989), 81.

[37]  George Victor, *Hitler: The Pathology of Evil* (Dulles, VA: Potomac Books, 1999), 136.

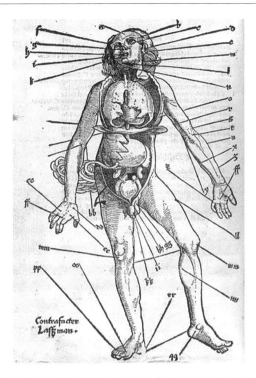

Figure 1. "Bloodletting Man" (Aderlassmann). Hans von Gersdorff, *Feldtbüch der Wundartzney* (Strasbourg: H. Schotten, 1528). Diagram indicating which blood vessels to bleed according to sickness. Courtesy of National Library of Medicine, Bethesda, MD.

phlegm, choler, bile, and blood.[38] Illness was often perceived as an imbalance in these elements, and doctors bled patients for every ailment imaginable. In an attempt to restore equilibrium, blood was let for pneumonia, fevers, back pain, diseases of the liver and spleen, rheumatism, and for a nonspecific condition referred to as "going into a decline."[39] The procedure was even performed as a prophylactic against illness. (See Figure 1.) In this period, physicians would occasionally experiment with the opposite, with giving blood instead of taking it away. Up to the seventeenth century, however, any therapeutic efforts to share blood were mostly ineffective, as the blood was orally ingested.[40] This was followed later by a primitive means

---

[38]   Starr, *Blood*, 7.
[39]   Ibid., 17.
[40]   Louis K. Diamond, "A History of Blood Transfusion," in Maxwell M. Wintrobe, ed., *Blood, Pure and Eloquent* (New York: McGraw-Hill Book Company, 1980), 660.

of injection of the donated blood, which often proved lethal owing to the ignorance of blood type compatibility.

From its beginning and regardless of its application—whether through ingestion or injection—the act of receiving blood elicited curiosity and fear. Apprehension regarding this new technique, which diverged from a centuries-old tradition of bleeding, was understandable. To obscure matters further, early blood transfusions were not always limited to life-or-death situations. As it was often the case in early modern Europe, some were conducted in the hope that they would alter personality traits. For instance, blood believed to contain the "ill-temperament" of a man would be replaced with that of a calmer animal, such as a lamb. In the words of British diarist Samuel Pepys, this procedure would give the once-irritable recipient "pretty wishes, as of the blood of a Quaker to be let into an Archbishop, and such like."[41]

The idea that blood contained the attributes of the creature came from (human or animal) became most apparent when clinical experiments with transfusing blood became more common. Robert Boyle, a British chemist and physicist in the seventeenth century, wondered whether a recipient dog would recognize his master, whether a dog transfused with sheep's blood would grow horns or wool, whether a small dog would change in stature if transfused with blood from a larger animal, and even whether marital discord could be treated by reciprocal transfusions of husband and wife.[42] Interpretations of this new technique were evident in popular discourse of the time. During the French Revolution, one man condemned to the guillotine wrote the National Assembly, begging it to take his blood, which would no longer be of any use to him, and transfuse it into the veins of an older man to see if rejuvenation would result.[43] The English play *The Virtuoso*, a satire on contemporary science, includes a scene in which there is a discussion of a cross-blood transfusion between a mangy spaniel and a healthy bulldog; predictably, this results in the spaniel becoming a bulldog,

---

[41] Samuel Pepys, a member of Parliament, is famous for his seventeenth-century diaries. Paul J. Schmidt and Paul M. Ness, "Hemotherapy: From Bloodletting Magic to Transfusion Medicine," *Transfusion* 46 (2006): 167.

[42] Raymond Hurt, *The History of Cardiothoracic Surgery from Early Times* (New York: Parthenon Publishing Group, 1996), 97.

[43] Ernest Flagg Henderson, *Symbol and Satire in the French Revolution* (New York: G.P. Putnam's Sons, 1912), 276.

and the bulldog a spaniel.[44] Though this may be simply a satirical remark on the mystical properties of blood, it accurately reflects the fact that early experiments in blood transfusion involved diverse research subjects. Blood might be exchanged between two different animals, or between one human and another, and not infrequently—as mentioned by Pepys—between humans and animals. In their efforts to realize possible beneficial properties of blood, researchers often did not confine themselves to one species when transfusing blood.

Occasional fears lingered that the recipient would then assume the mental or physical characteristics of the animal donor. On a much more practical note, provided the quantity was large enough, blood exchange between species did not result in human adaptation of animal traits, but instead was often fatal, as their blood was mutually incompatible. For the same reason, transfusions between humans were similarly perilous. Before the blood types were discovered, even receipt of human blood often produced adverse effects, including death. The repeated problematic outcome of blood transfusion led to its international prohibition in humans in the late seventeenth century.[45] Still, intermittent reports of blood exchange were scattered throughout the medical literature of the next 150 years.[46] Medical progress in the mid-nineteenth century revived interest in the possibility of blood transfusion therapy. The first transfusion with human blood, for the purpose of replacing lost blood and not as psychotherapy, was probably made in the early 1800s by the British obstetrician James Blundell, who was frustrated by the loss of so many patients to postnatal hemorrhaging.[47]

Despite this advance, most efforts continued to be strikingly similar to those in the seventeenth century. Because the existence of blood types

---

[44]  Wintrobe, ed. *Blood, Pure and Eloquent,* note 663.

[45]  Jean-Baptiste Denis, a French mathematician, is often regarded as the first person to perform a blood transfusion on a human being. When his patient died from the procedure, Denis was placed on trial for murder. Although Denis was exonerated, the Paris Society of Physicians declared its opposition to such experiments and persuaded the Criminal Court in Paris on April 17, 1668, to forbid further transfusions without approval from the Faculty of Medicine of Paris. This affected transfusion therapy outside of France as well. See Wintrobe, ed., *Blood, Pure and Eloquent,* 665. See also Pete Moore, *Blood and Justice: The Seventeenth-Century Parisian Doctor Who Made Blood Transfusion History* (Chichester, England: John Wiley and Sons, 2002).

[46]  Ibid., note 665.

[47]  "John Henry Leacock published his dissertation in medicine at the University of Edinburgh in 1816. In it he proposed the transfusion of human blood as treatment for both a "deficiency" in the blood and a loss of blood. Leacock performed animal experiments proving that therapeutic transfusion was possible if species specificity of donor and recipient was matched. He was the first to define transfusion as medical therapy." Schmidt and Ness, "Hemotherapy," 167.

was still unknown, transfusions remained, at best, unpredictable. In 1875 German physiologist Leonard Landois (1837–1902) published statistics revealing the discouraging situation of blood group research at this point: of 129 transfusions conducted on humans (predominantly with sheep blood), sixty-two were lethal.[48] This led him to caution against transfusing "foreign," or non-human, blood into humans.[49] It was still practiced, however; separate statistics published the same year similarly reported that 48 percent of more than 100 blood transfusions, mainly with blood from lambs, ended fatally.[50] Landois' advice did not greatly help matters; of 347 human-to-human exchanges in 1875, 180 had "unfavorable results."[51] In another report, fifty-two of 359 such transfusions had taken a "less fortunate ending."[52] Additional efforts to substitute organic substances instead of blood—such as milk, peptone, gelatin, serum, or albumin—often only worsened matters.[53] Transfusion in general was still not even a concern for many; between 1897 and 1900, it was not mentioned even once in Germany's respected medical publication *Zentralblatt für Chirurgie* (Surgical Leaflet).[54] In spite of other, impressive medical advancements in the nineteenth century, the unreliability of transfusion thwarted progress in serology. This was apparent not only in individual cited statistics, such as those of Landois, but also in general medical literature.

Thus the reference in *The Operated Jew* to the fact that blood transfusions had fallen "out of fashion" is accurate. As a medical doctor, Panizza certainly would have been aware of this. Perhaps to exploit apprehensions concerning blood exchange, Panizza includes the transfusion episode to

---

[48] D. Wiebecke et al., "Zur Geschichte der Transfusionmedizin in der ersten Hälfte des 20 Jahrhunderts (unter besonderer Berücksichtigung ihrer Entwicklung in Deutschland)," *Transfusion Medicine and Hemotherapy* 31 (2004): 12.

[49] Ibid.

[50] Ibid.

[51] Ibid. "In 1873, Gesellius presented similar statistics showing that 146 out of 263 transfusions (56 percent) had fatal consequences." Army Medical Research Laboratory, *Selected Contributions to the Literature of Blood Groups and Immunology IV, Part II: Blood Groups and Their Areas of Application* (Fort Knox, KY: January 1971), 64–65.

[52] A. W. Bauer, "From Blood Transfusion to Haemotherapy—The Anniversary of the German Society for Transfusion Medicine and Immunology (DGTI) from a Medicinal-Historical and Bioethical Perspective," *Transfusion Medicine and Hemotherapy* 31 (2004): 4. Similar developments were occurring in the United States; "syringe transfusion was documented by the Union armies in the Civil War, and blood transfusion was photographed at Bellevue Hospital in New York in 1873." Schmidt and Ness, "Hemotherapy," 167.

[53] Wiebecke et al., "Zur Geschichte der Transfusionmedizin," 12.

[54] Ibid.

address persistent ideas that an exchange of blood could alter one's physical or mental state; as late as 1868, one German professor of medicine claimed that a girl would assume the characteristics of a cat after having received cat's blood.[55] The developing race science of the nineteenth century incorporated pre-modern concepts of race. The implication remained that an individual's personality was in his or her blood, and to this was added his or her "racial type." *The Operated Jew* demonstrates this shift in emphasis, even though there was still no recognizable difference between the blood of one individual—or even species for that matter—and another. At the same time, Panizza's inclusion of the procedure of blood transfusion pointed to increasing trends in experimental research of the time.

## ORIGINS OF SEROLOGY

In the late nineteenth century, shortly after Panizza's book was published, quantifiable differences were found to exist within blood. By this point in time, most physicians who dabbled in blood research would have been aware that the blood of one individual, or animal, would occasionally cause that of another to "clump" when mixed. What researchers did not know was that this was a normal reaction between what were incompatible blood types. Instead, early blood scientists thought that this clumping stemmed from some type of pathology, or sickness, in one of the subjects. Accordingly, analyses of blood often involved drawing blood from diseased subjects. In 1900, while conducting research on blood and disease at the Institute for Pathological Anatomy at the University of Vienna, Austrian physician Karl Landsteiner (1868–1943) discovered the blood groups.[56] Landsteiner noticed a systematic pattern in interactions of different blood samples, and he attributed this not to pathology, but to simple individual differences in the blood. He confirmed this theory through further analyses—there was, in fact, the "unexpected existence of clearly demonstrable

---

[55] Fritz Schiff claimed that Dr. Irenaeus Vehr believed this in 1868. See Army Medical Research Laboratory, *Selected Contributions*, 60.

[56] "Blood clumping was first recognized by Landois Eulenberg in 1866, when he demonstrated that the serum of one animal may have the property of destroying the red cells of another animal of the same species." J.A. Buchanan and E.T. Higley, "The Relationship of Blood Groups to Disease," *British Journal of Experimental Pathology* 2 (1921): 247.

differences between the bloods within one species."[57] Because the sample size was relatively small, no subject tested had the rarest blood type, AB. Landsteiner proposed that there were only three different blood groups, which he referred to as groups O, A, and B.[58] Due to the predictable differences, Landsteiner suggested that his findings might be of use in transfusing blood; however, he did not recognize the importance of his contribution and merely commented that he hoped it would be "of some use to mankind."[59] In 1901 his theory was quietly published in the *Wiener Klinische Wochenschrift* (Viennese Clinical Weekly).[60] In part because of the novelty of his claim, and the slow spread of news in the less interconnected world of the early twentieth century, it would take many years for his work to be recognized. In fact, the breakthrough received so little attention that others who later made the same findings believed they were the first to "discover" the blood types. Unaware of Landsteiner's findings, Czech physician Jan Jansky also reported the existence of different blood types, but in 1907, and he labeled them I, II, and III.[61] Three years after Jansky, American William Moss did the same, but instead named them IV, II, and III respectively.[62] As a result, until an international nomenclature was established in 1927, the blood types were confusingly referred

[57]   A. D. Farr, "Blood Group Serology—The First Four Decades (1900–1939)," *Medical History* 23 (1979): 215. In 1899 an English physician, S. Shattock, also observed in his patients the phenomenon of isohemo-agglutination—the agglutination (clumping) of bloods when mixed. See A. Furukawa, "A Study of Temperament and Blood Groups," *The Journal of Social Psychology* 1 (1930): 494–508. The fact that Landsteiner initially suspected that agglutination stemmed from a pathological condition speaks to the scientific perspective of modern medicine at this time. Landsteiner unwittingly reinforced associations between "blood mixing" and infection that were used in race propaganda. Spörri, "'Reines Blut,' 'gemischtes Blut,'" 216.

[58]   "Landsteiner's original labeling of the blood types is still that used today. Landsteiner was stung severely by subsequent American scientists' criticism for having only discovered three blood groups. As a result, he wrote a letter to Adriano Sturli, in which he asked the latter to confirm in writing that he (Sturli) was Lansteiner's pupil and that he had collaborated with Alfred von Decastelo to perform supplementary work in this field. The discovery of the fourth blood group by Decastelo and Sturli, performed on a much larger sampling, was initially designated as "of no particular type." However, it was later designated as blood group AB". A. Matthew Gottlieb, "Karl Landsteiner, the Melancholy Genius: His Time and His Colleagues, 1868–1943," *Transfusion Medicine Reviews* 12, no. 1 (1998): 18–27.

[59]   Ibid., 20. Theoretically, considering the unreliable nature of blood exchange at this point, this new research should have had immediate medical consequences. For the first time, it offered the possibility of making a suitable donor choice before starting the transfusion and documentation for the methodical construction of serological transfusion theory. See Farr, "Blood Group Serology," 215.

[60]   In Karl Landsteiner, "Über Agglutinationserscheinungen normalen menschlichen Blutes," *Wiener Klinische Wochenschrift* 14 (1901): 1132–1134.

[61]   Farr, "Blood Group Serology," 217.

[62]   Ibid.

to in double- or even triple-meaning, depending on whose system was being used.[63]

It was nearly a decade after Landsteiner's 1901 discovery that doctors Emil von Dungern and Ludwik Hirszfeld outlined the potential anthropological significance of his work. Hirszfeld, born a Polish Jew in Warsaw in 1884, received his medical doctorate from the University of Berlin in 1907.[64] Afterwards he assisted in the serological department of the Institute for Cancer Research in Heidelberg for its director, von Dungern. Suspecting that blood type was an inherited trait—something that had not yet been determined—von Dungern and Hirszfeld drew and recorded the blood types of seventy-two Heidelberg families, which enabled them to partially record three generations.[65] The blood types revealed a definite pattern of heredity; for example, a child of type B with a mother of type A could never have a type A or O father; and a child of type O could never have a type AB father (regardless of the mother's type).[66] They concluded that blood type was a permanent biological feature that was passed from parent to child and therefore followed Mendelian Law.

Von Dungern and Hirszfeld's research was presented to the medical community at a particularly opportune time. Interest in Mendel's laws revived around 1900, and anthropological studies, as a result, increasingly focused on traits that were known to be passed from one generation to the next. Since its origins in the nineteenth century, analysis of inherited traits, whether normal or pathological, was the conventional means of racial classification. Because of this, von Dungern and Hirszfeld theorized that blood types might offer a new method of racial analysis; they thus became the first physicians to suggest an explicit, scientific link between blood and race. Testing this theory, though, would be a daunting task. At seventy-two families, the subject group pooled to examine the heredity of blood type was quite extensive, but determining any relationship between blood and race would require research on a much larger scale. A sufficient study would necessitate not only traveling to widely disparate regions to examine

---

[63]  Ibid.

[64]  Leo J. McCarthy and Mathias Okroi, "The Original Blood Group Pioneers...The Hirszfelds," *Blood Banking and Transfusion Medicine* 2, no. 1 (2004): 25.

[65]  Paul Steffan, ed., *Handbuch der Blutgruppenkunde* (Munich: J.F. Lehmann, 1932), 5.

[66]  Frank Heynick, *Jews and Medicine: An Epic Saga* (Hoboken, NJ: Ktav Publishing House, 2002), 434.

various ethnicities, but also collecting enough samples to be representative of the group as a whole. Aware of the time-consuming, labor-intensive, and expensive nature of such an undertaking, von Dungern and Hirszfeld admitted that their theory would likely remain just that. As Hirszfeld would later describe it, they anticipated "a project which, under normal circumstances, would have required dozens of years of work: a project of world exploration from the serological point of view. It is possible that the project would not have been realized in our lives."[67]

## THE *VÖLKISCH* NOTION OF "BLOOD DEFILEMENT"

Despite the fact that no such study linking blood and race had yet been conducted, propaganda of the extreme right continued to suggest there was a definitive medical relationship between the two. Perhaps most notable among such works is Artur Dinter's *The Sin against the Blood* (*Die Sünde wider Das Blute*). Published in 1917, Dinter's novel quickly went on to be purchased by nearly a million readers in the interwar period, thus becoming one of the most successful racial propaganda works of the Weimar Republic. His work employs the conventional physiognomic racial stereotypes but, like Panizza, also draws upon the unseen racial trait of blood. *The Sin against the Blood* centers on the German protagonist, Hermann Kämpfer, portrayed as the "Aryan ideal" with his blond hair, blue eyes, and racially pure, agricultural background. Nevertheless, both of the children he has with his similarly "Aryan" wife Elisabeth Burghamer have an appearance that clearly suggests Jewish descent. The first son is described as "dark-haired with dark skin, and covered in curly dark hair." The second child also appears to be a "Jewish lad" (*jüdisch Knabe*). The reader comes to learn that the source of their "racially inferior" traits is their mother's racial makeup. Though Elisabeth is herself the daughter of a German and a Jew, the negative influence of her "Jewish blood" takes precedence in their offspring. Later, Kämpfer fathers another child with a second wife, Johanna, whose blond hair and blue eyes again serve as definitive indicators of "Aryan lineage." However, this third child, too, appears to be "authentically Jewish."

---

[67] William H. Schneider, "Chance and Social Setting in the Application of the Discovery of Blood Groups," *Bulletin of the History of Medicine* 57 (1983): 556.

Johanna may look "Aryan," but Dinter explains how her blood has been unknowingly, permanently defiled by a single instance of intercourse with a Jew. While Dinter fails to detail the biological mechanics of this change, the implication is that Johanna absorbed the Jew's semen, which then permanently altered her racial makeup.

In contrast to Panizza's fictional account, Dinter does not refer specifically to clinical procedures. His focus is defilement of blood not through transfusion, but through "interracial" sex.[68] Like Panizza's book, Dinter's work has medical overtones, but they are now sexual as well. *The Operated Jew* includes exchange of blood as a critical step in racial transformation. Panizza stresses the clinical aspect of this exchange. The Jewish character Stern, a medical student, first made the necessary incision himself. The eight liters of German blood accurately corresponds to the approximate blood volume in a human body. Panizza weaves indistinct notions of blood and race with medical objectivity. Dinter also maintains that race is contained in the blood but is subject to change through miscegenation. The biological change is implied, but Dinter foregoes any parallels with modern medicine, as these were conspicuously absent.

Increased sexual contact between Germans and Jews—unmistakable in the rising frequency of intermarriage at this time—was seen by anti-Semites as a serious threat to German identity. In spite of their differences, both Panizza and Dinter demonstrate the malleable nature of blood and race propaganda. Although the acts are different—one is a blood transfusion, the other sexual intercourse—the end result is the same. Racist propaganda was more inclined to emphasize the threat of blood defilement from miscegenation, as this was much more common than blood transfusions, which were exceedingly rare. Long before the Third Reich, then, a pattern was set in which blood and race could interact in either a clinical or sexual context.

Importantly, both Panizza and Dinter's works imply a clinical reality between blood and race without abandoning physiognomic "racial indicators," as these continued to be the mainstay of racial classification. The racial satire, *Der Operierte Goy* (The operated Goy), published in 1922, fol-

---

68   "The term 'miscegenation' was first coined in the United States in 1863 to describe 'interbreeding between Anglos and non-Anglos.'" Stefanie Wickstrom, "The Politics of Forbidden Liaisons: Civilization, Miscegenation, and Other Perversions," *Frontiers: A Journal of Women Studies* 26, no. 3 (2005): 168–198.

lows this same pattern. This work centers on an anti-Semitic German family, the von Reshoks. Their well-documented racial purity was unmistakable in their "perfectly Aryan" appearance:

> They all had thin lips, Prussian chins, proud necks, and fabulously slender builds, and their legs, which in their innocence did not know either X or O, stood simultaneously on aristocratic and pan-Germanic feet and took strides as though descending from Mount Olympus. Above all, they did not have narrow eyes with a dark brown gloss but open, true blue ones which glistened like pure ice.[69]

In order to "keep his blood pure," the von Reshoks' son, the single Count Kreuzwendedich von Rehsok, is cautioned to be careful in selecting a spouse. However, he is seduced by a wealthy, attractive Jewess, Rebecca Gold-Isaac. Protecting their blood was also a concern for the Isaacs, as they had "never been contaminated by alien [non-Jewish] blood."[70] Rebecca wants her beau to atone for his family's anti-Semitism; she demands that he physically and spiritually renounce his Aryanism and become Jewish. Like Itzig Faitel Stern in *The Operated Jew*, Count von Rehsok seeks the help of various medical experts to complete his racial transformation. He visits Dr. Friedlander, a prominent orthopedist, in order to "convulse his Aryan body" into a Jewish one:

> The doctor devoted special attention to the nose, which he endowed with an artificial hump and made the tip curl over. Thereafter, he performed one of his most famous spinal atrophies. The count's bones were broken at their joints and then carefully brought to heal in the shape of an egg. Then he disappeared, with brand new flat feet, to Romania to learn Yiddish from the rabbis there, as well as the gestures that go along with it.[71]

The procedures are a success, resulting in von Rehsok's change "from a German into a Jew." Brief reference is made to the "fact" that Rebecca's marriage to von Rehsok would contaminate her own blood. This is portrayed as negligible compared to the "de-Nordification" (*Entnordung*) of the

---

[69]  Zipes, *The Operated Jew*, 76.
[70]  Ibid., 79.
[71]  Ibid., 83–84.

German race. Curiously, *The Operated Goy* suggests that racial defilement is not gender-neutral. Von Rehsok is portrayed as losing his own blood purity through a successful operation; however, the male Jew in *The Operated Jew* fails in his attempted Aryanization operation, even though the procedures are essentially the same. The notion that race therefore trumps gender, and that Jewish traits "stick" more than Gentile ones, is an underlying presumption of this racist genre. Von Rehsok's own blood is affected differently from that of the Aryan wife in Dinter's novel, with the common result that Jewish blood proves more immutable than gentile blood.

Anti-Semitic descriptions of blood defilement were predominantly concerned with the possible negative consequences of miscegenation for "Aryan" women. As demonstrated in Dinter's novel, sexual intercourse, whether it resulted in children or not, contaminated the blood of the racially superior individual. Even if she had only one sexual encounter with a man of a "less worthy race," an "Aryan woman" then became unfit for "noble breeding."[72] Comparable propaganda describing the threat from sexual intercourse between German men and Jewish women does not seem to exist. A different set of concerns was applied towards German women because of their duty to bear members of the next "Aryan generation." After sexual contact with a "racial other," a German woman could then only conceive racially inferior children—even if their biological father was "Aryan." These supposed negative results of race mixing, which provide the storyline for *The Sin against the Blood* and other racial diatribes, were by no means unique to Dinter and were not limited to the realm of anti-Semitic fiction. Rudolf John Gorsleben (1883–1930), who began a postwar Aryan occultist movement, railed against the vulgar, corrupt, and wretched modern world as the "sad result of racial mixing."[73] Just as in *The Sin Against the Blood* and *The Operated Goy*, Gorsleben believed blood defilement applied differently to men and women. Miscegenation was detrimental for the racially superior man, since his purity was "debased in the progeny"—in other words, not in himself.[74] An Aryan woman, on the other hand, could be "impregnated" by intercourse, even when no con-

---

[72]   Lauper, ed., *Transfusionen*, 35.
[73]   Nicholas Goodrick-Clarke, *The Occult Roots of Nazism: Secret Aryan Cults and Their Influence on Nazi Ideology* (New York: New York University Press, 1993), 157.
[74]   Ibid.

ception occurred, so that her subsequent offspring would bear the characteristics of this (prior) lover.[75]

These works illustrate a larger trend within anti-Semitic fiction before 1933 to rely upon a discourse of blood. Each example either uses blood rhetoric (Dinter) or plays upon its use in racist propaganda (Panizza, Mynona). The primary concern of racially biased German anthropologists was that racial mixing would lead to the degeneration of both participants—though particular emphasis was placed upon the sexual contamination of the female German constituent, instead of vice versa. Each author framed his narrative in the medical language of clinical procedures, as each recognized the advantage of incorporating eugenic notions into popular literature, which then helped to legitimize racism as scientific fact. Aware of its usefulness, far-right propagandists would continue to employ blood rhetoric well into the Third Reich.

On occasion, as circulating notions of blood and race intensified, there were responses to distortion of medical fact during the liberal Weimar Republic. In 1921 writer Hans Reimann published a parody of Dinter's work entitled *Die Dinte wider das Blut, ein Zeitroman* (A mark against the blood: A novel), in which he blatantly mocked the medical propaganda of the far right.[76] In it, Reimann's German protagonist, Professor Hermann Stänker, conducts research on the blood of Germans and Jews. He notes that, under a microscope, the German blood is characterized by "Aryan blood cells," while the Jewish blood is made up of so-called "Semitokokken." Upon placing the two different bloods in a laboratory flask together, Stänker expects the "Israeli bacteria" to be destroyed by the "Teutonic blood."[77] In the presence of the "Aryan blood," he reports, the "half-moon-shaped Semitokokken" press themselves into the corner of the flask and shake with fear; it seems that the Aryan blood has "gained the upper hand!"[78] The Jewish blood cells are then carefully separated and placed in a vaccine. When they are injected into a rabbit, the animal immediately gestures submissively with its front paw and freezes—clearly, Stänker observed, the crea-

---

[75] Ibid.

[76] Artur Sünder, *Die Dinte Wider das Blut, ein Zeitroman* (Hanover: Paul Steggemann, 1921). Artur Sünder was one of the numerous pen names used by Hans Reimann. See also Alan Levenson, "The Problematics of Philosemitic Fiction," *German Quarterly* 75, no. 4 (2002): 379–393.

[77] Sünder, *Die Dinte Wider das Blut*, 8.

[78] Ibid., 13.

ture was in the trance of a "foreign racial power."[79] With a pen and pencil, the rabbit proceeds to solve a difficult mathematical equation. When asked its name, it responds by writing "Baruch Veilchenblüth," and explains that it is from Köln.[80] The scientist further observes that the rabbit now has an "authentic Jewish mark." After returning from a brief trip, Stänker discovers that among his thousands of laboratory rabbits, many have become unexpectedly "Judaized" (*verjudete*) in his absence.[81] The Jewish rabbits spread their tainted blood throughout the remaining population. To counteract this process, Stänker reasons that either an anti-Semitic antidote has to be found, or the "Aryan blood" has to be thoroughly saturated with "national consciousness." Otherwise, he fears, the now-mixed blood of the "racial mongrel" will triumph.[82]

The racial propaganda of the late nineteenth century represents a sea change towards "scientific" anti-Semitism, which benefited from a centuries-old base of Christian anti-Semitism. Medicine was favored because science was regarded as autonomous from all other forms of thought (i.e., religious, moral, or political). Of course, loose references to differences in blood and race continued to be used freely after the introduction of race science but were increasingly supported by the idea that race was biologically in the blood.

Repeated allusions to differences between the bloods of individuals were partially realized in Karl Landsteiner's discovery of the blood groups. Why did racist propagandists not make use of this finding? Landsteiner's work was useful in labeling the blood types of individuals, but this did not have any direct appeal for scientific or popular racial theorists. There were several reasons for their lack of interest. First, Landsteiner came across the existence of the blood groups by analyzing a relatively small subject group (only 105 of his colleagues), all of whom were Austrian. He did not record any other characteristics in his analysis—such as pigmentation, skull shape, or place of birth—and therefore did not attempt to relate blood type to any other conventional (racial) characteristic. For the time being, the lim-

---

[79]  Ibid., 14. Laboratory animals were common in serological research. Blood type differentiation between species was examined even before World War I in rabbits, but also cattle, chickens, dogs, and pigs. See Steffan, ed., *Handbuch*, 7.

[80]  "Baruch," Hebrew for "blessed," was a common given name among Jews, as was the surname *Veilchen* (Violet).

[81]  Sünder, *Die Dinte Wider das Blut*, 21.

[82]  Ibid., 22.

ited number of subjects, combined with their apparent racial homogeny, prevented any larger presumptions about the blood types and racial classification. From an anthropological perspective, the newly discovered differences in blood were basically useless—they made no cultural or ethnic distinctions possible.[83] What was needed, as von Dungern and Hirszfeld already realized, was an extensive study on diverse peoples.

## SEROANTHROPOLOGY

Several articles were published between 1916 and 1921 on the distributions of blood types.[84] As was the case with the discovery of the blood types, there was a disconnect in communication in the international medical community over developments in the relatively new pursuit of serology. The first seroanthropological studies were conducted independently, each acting upon the idea that blood type was a heritable, and therefore possibly also a racial, trait. While their purpose was to analyze blood type and race, their results did not elicit interest among race theorists. Instead, the studies were criticized for the small size and the homogeneity of their subject groups (only Europeans or those of European descent). Anthropologists knew that determining the racial value of blood-type patterns required that representative groups of the different races be examined, and in adequate numbers. A unique set of circumstances during World War I would provide these necessary criteria. The war affected all aspects of European society, including medicine. Redirecting of funds and personnel frequently interrupted progress at home, but wartime changes also presented scientific research opportunities.

When World War I broke out, Serbia was devastated by epidemics of typhus and bacillary dysentery. In 1915 Ludwik Hirszfeld applied for duty there, ostensibly for medical reasons but also in support of pan-Slavic ideas

---

[83] Lauper, ed. *Transfusionen*, 217.

[84] "These studies were Hara and Kobayashi, 353 Japanese subjects tested in Nagano, results published in 1916 in *Jji-Shinbun*; Kilgore and Liu, 100 Chinese subjects tested after transfusion accident, 1918, *Chinese Medical Journal*; Hirszfeld and Hirszfeld, 7,900 subjects tested, 1919, *Lancet* and *Anthropologie*; Moffitt et. al., 1,122 subjects tested at U.S. army base, 1919, *Journal of the American Medical Association*; Weszeczky, 550 Hungarians, Romanians, and Germans in central Hungary, 1920, *Biochemischer Zeitschrift*; Alexander, fifty cases of "malignant" and infectious diseases, published in 1921 in *British Journal of Experimental Pathology*; Culpepper and Abelson, 5,000 tests at Parke Davis (Detroit), published in 1921 in *Journal of Laboratory and Clinical Medicine*." See Schneider, "The History of Research on Blood Group Genetics," Table 1, 283.

(Hirszfeld was Polish). Serving as serological and bacteriological advisor, he would remain with the Serbian army until the end of the war. Both Ludwik and his wife Hanna, a pediatrician, would spend their first two years of service in Serbia.[85] In 1916 a German offensive in Belgrade forced their evacuation to Salonika, where the Serbs were soon joined by an Allied expeditionary force.[86] Unable to advance militarily, but politically too embarrassed to evacuate, the Allies were trapped in "the cage" (as they themselves called it), or what the Germans referred to as "our largest POW-camp."[87] For the next couple of years, between 1916 and the end of hostilities in 1918, an entire contingent of Allied troops was confined to the area in and around Salonika.

Hirszfeld recalled the unlikely prospect he and von Dungern had outlined years earlier. He realized that he was in an exceptional position to examine a possible relation between blood type and race. Ironically, the Allies' entrapment in Salonika fulfilled the requirements perfectly. The Allied troops were made up of many different nationalities, including groups from throughout Europe, as well as "natives" from various European colonial settings, and there were sufficient numbers of each "racial type." Furthermore, the subjects were not related to one another, thereby negating any possibility of hereditary distribution. The study would be relatively easy, as determining blood type was a simple procedure; the examiner only had to prick the individual's fingertip and wait half an hour to observe the reaction. The Hirszfelds quickly began their research. Over the course of their isolated two-year stint in Salonika, they were able to collect the blood types of between 500 and 1,000 representatives of sixteen different races. In the summer of 1918, their results were first presented to the Salonika Medical Society in an article entitled "Serological Differences between the Blood of Different Races: The Results of Researches on the Macedonian Front." The survey fulfilled the expectations Hirszfeld and his mentor voiced years earlier; the patterns suggested some connection between blood type and race.

In addition, the study confirmed Hirszfeld's theory that an individual's serological type did not change over time, as at the time of his analysis, he

---

[85] During his time, serological in the hospital for contagious diseases in Salonika, Hirszfeld discovered the bacillus "Salmonella paratyphi" C, today called "Salmonella hirszfeldi."

[86] The Allied expeditionary force was retreating from the defeat at Gallipoli.

[87] Heynick, *Jews and Medicine*, 435.

observed that the blood types were the same as they had been eight years earlier (he had typed his colleagues' blood before the war)—despite the fact that one of them had had typhoid and another had been suffering from chronic malaria.[88] Nor did type seem to be affected by disease or environment. By 1918, after decades of attempts to pinpoint a definitive racial indicator, racial anthropologists would have been quite receptive to the prospect of a new, heritable (and therefore fixed) trait for racial analysis. Physiognomic characteristics were the "go to" variables in racial categorization, but by the early twentieth century they were under increasing scrutiny, mainly because of the possibility that they changed with environment. German anthropologist Franz Boas's landmark study of 1914 certainly suggested as much. Boas and his assistants measured the skull shape and the stature of the parents and offspring of four American immigrant groups: Central European "Slavs"(Bohemians, Poles, and Hungarians), Jews, Sicilians, and Neapolitans.[89] Their findings revealed that the physiognomy of the children differed from that of their parents. Apparently, very "round-skulled" Jews were becoming "long-skulled," and the very long skulls of Sicilians were becoming rounder over time. These results indicated that physical features could change relatively rapidly in response to changes in environment and nutrition. The prospect that surroundings could enact a difference in appearance would obviously render physiognomic racial classification ineffective.[90] The

---

[88] Ludwik and Hanna Hirszfeld, "Serological Differences between the Blood of Different Races: The Results of Researches on the Macedonian Front," *Lancet* 2 (1919): 675–679. Though the existence of the blood groups was generally recognized, the Hirszfeld study's detailed inclusion of those sick and healthy subjects reveals that the agglutination between differing blood types was not yet completely recognized as normal. It would take years for the medical disciplines to thoroughly realize that agglutination was a normal reaction between incompatible blood types and not the result of some other factor, such as a bacterial infection. The Hirszfelds carefully noted the health of those examined; however, they never observed any difference in blood-type distributions between healthy and sick groups. Ibid., 677.

[89] Exactly which Central or Eastern European peoples were classified as "Slavs" was subject to debate. For instance, one German anthropologist categorized the following Europeans as Slavic: East Germans, Finns, Hungarians, Poles, Lithuanians, and Russians. Still others claimed that the Hungarians were "Magyars," not Slavs, and many peoples in Finland, Poland, and parts of Russia were "Nordic." Even with such discrepancies, there were countries generally considered Slavic, namely Russia, Poland, and Bohemia. See Siegmund Wellisch, "Blutsverwandtschaft der Völker und Rassen," *Zeitschrift für Rassenphysiologie* 1, 1928: 21–34, 30 and Emily Greene Balch, *Our Fellow Slavic Citizens* (New York: Charities Publication Committee, 1910), 6.

[90] Franz Boas, "Changes in the Bodily Form of Descendants of Immigrants," *American Anthropologist* 14, no. 3 (July-September, 1912): 530–562. While he examined the hair color of his subjects, Boas did not report the popular racial criteria of eye and skin color. Eye color was excluded as, he explained, it was subject to "very strong personal equations," and exposure to "air and light" could cause considerable variations in the classification of skin color. Simply put, Boas believed each to be too subjective. Sophisticated for its time, Boas's study showed statistically that given different circumstances, people defined as belonging to a given "racial type" could be expected

results were so convincing, however, that Boas went from defending to criticizing the coveted cephalic index as a means of racial analysis.

By contrast, the Hirszfelds' study indicated that blood type was not subject to change, regardless of environment. The subjects examined shared many of the same circumstances—they were predominantly soldiers, who had been living in the same climate, exposed to (or having suffered from) the same diseases, and eating similar food (with the exception, Hirszfeld noted, of the Indians, who were generally vegetarian). Even after exposure to the same conditions over the course of two years, the differences in blood type remained. This was corroborated by the blood-type distributions of the minority groups examined. For instance, Hirszfeld observed that the blood-type frequencies of the Jews in Monastir, who could trace their ancestry in the area back several centuries, differed from those of the surrounding Christian majority. Despite their extended coexistence, each group had a distinct blood type profile.

As with any scientific study, the Hirszfeld study provided thorough details on the proportions of those examined and the technique employed. Blood samples were drawn from doctors, medical assistants, patients in military hospitals, prisoners, soldiers, schoolchildren, and civil servants. Although the four blood types were found among all of the ethnicities studied, the Hirszfelds ultimately chose to focus solely on types A and B. Their reasons for doing so were twofold: von Dungern and Hirszfeld believed A and B to be dominant hereditary characteristics, and these types demonstrated the most conspicuous differences among regional populations in the wartime analysis. The geographic origins of the subjects yielded a definite pattern: Blood type A appeared more common in Northern and Western Europe. As one moved south and east, this was gradually replaced by type B. The blood types of the Balkan peoples examined, including Greeks, Bulgarians, and Serbs, followed this trend.[91] The Russians also had more

---

to physically change over time. See Gretchen E. Schafft, *From Racism to Genocide: Anthropology in the Third Reich* (Champaign: University of Illinois Press, 2004), 207. See also John S. Allen, "Franz Boas's Physical Anthropology: The Critique of Racial Formalism Revisited," *Current Anthropology* 30 (1989): 79–84.

[91]  Of the 500 Greeks examined, 300 came from Old Greece and the Islands, and an additional 200 refugees tested came from Asia Minor and Thrace. The 500 Bulgarians tested, all healthy prisoners, yielded results absolutely identical to those of the Greeks and Serbs. The Turkish sample of 500 "Mohammedans" (Muslims), composed of 150 schoolchildren from Salonika, 150 civil servants from Micra, and 200 civil servants from Macedonia, was also found to diverge from the European type. See L. and H. Hirszfeld, "Des méthodes sérologiques au problème des races," *L'Anthropologie* 29 (1918–1919): 526.

type B and less A than the Europeans of northwestern Europe. The same applied to the "Arabs" from Tunisia and Algeria and the "Negro soldiers" from Senegal. The highest frequencies of type B were reported among those from the regions farthest east—the Indochinese and the "Hindus" of South Asia. Consequently, the Hirszfelds affiliated type B with Eastern peoples and attributed its distribution farther west to the immigration of "Asiatic peoples."

After the Hirszfelds compiled their results in a diagram—with type A represented by a white line and type B with solid black—the higher percentage of type A among northwest Europeans was unmistakable, a pattern the Hirszfelds referred to as "remarkable."[92] The varying blood-group distributions of those in Salonika implied that blood might be a useful means of racial analysis.[93] Unlike physiognomy, however, which could be used in the anthropological classification of an individual, the racial significance of blood was contingent upon group proportions. The majority of Western Europeans tested did not have less than 45 percent type A blood. The decline in type B became more obvious as one moved west; the Germans were about 17 percent, the French and Italians 14 percent, and the English merely 10 percent.[94] The reverse was the case in Africa and Asia: here, type A averaged only about 27 percent.[95] To articulate this pattern, the Hirszfelds formulated what they referred to as the "biochemical race index," an equation which compared the incidence of blood types A and B within a group. The resulting figure categorized the group as one of three different racial types. A high enough incidence of type A blood would yield an index greater than two, which was characteristic of "Western European" types, and included peoples spanning the approximate geographic region from Greece to England. The opposite, with an index of less than one, was the "Asiatic-African" category. The Hirszfelds labeled the groups in between one and two as "intermediate"; this included mainly the people of the Mediterranean basin, considered a transitional region between Western Europe and Asia/Africa, but also included peoples of the Middle East, Jews, and some groups in Russia. The blood group distributions of the "Slavs, Serbs,

[92] Ibid.
[93] Ibid., 677.
[94] Ibid., 678.
[95] Ibid.

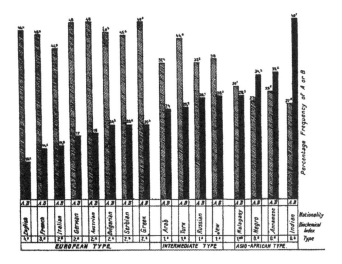

Figure 2. Graph of the Hirszfelds' biochemical race index. The proportion of
type A to type B blood determined whether a group was "European,
Intermediate, or African" in type. L. and H. Hirszfeld, "Serological Differences
between the Blood of Different Races: The Results of Researches
on the Macedonian Front," *Lancet* 197, no. 2 (1919): 675–679

and Bulgarians" placed them at the "extreme limit of the European type."[96]
Central Russia and Siberia showed the "intermediate type," while Russians
in the Ukraine and Volga basin were classified as "Asiatic."[97] The Hirszfelds
attributed the blood type distributions of Russians to "frequent nomadic
invasions" from central Asia. (See Figure 2.)

Based on the shift in distributions between East and West, the Hirszfelds
reasoned that it "would be very difficult to imagine a single place of origin
for the human race." Instead of the human race emerging with one blood
type, the pattern of varying frequencies suggested that there had been two
centers of origin—one in India, the other in Europe. India, which had
the highest frequencies of type B blood, was referred to as "the cradle of
one part of humanity." The other point of origin was somewhere in Western Europe; its exact geographic location was not clear. Initially, prior to
the movements of different peoples across the Eurasian continent, the
Hirszfelds believed that A and B had likely been in the same proportions.
Over time, their distributions were gradually altered by "mutual diffusion,"

[96] Ibid., 537.
[97] Ibid.

which is to say that the early, unmixed types A and B had been diluted by generations of miscegenation between the two.[98] The highest concentrations of type A or B identified where the "purest" races were—before racial migration and the miscegenation that followed.

Unable to publish the results in enemy Germany, the Hirszfelds submitted the results of their research, including the biochemical race index and their theory regarding the separate origins of the human race, for publication in the *British Medical Journal*. Months later, the editor curtly replied that the study "would not be of interest to physicians," and the manuscript was returned.[99] Perhaps the most unfavorable result, which the Hirszfelds themselves were careful to point out, was the fact that the four blood groups were present in every group examined.[100] This eliminated the possibility of precise, individual racial classification through blood. Admittedly, there were patterns among racial groups and their blood-type frequencies, but they did not reveal a definitive correlation between blood and race—only frustrating proportions. Furthermore, the idea of two original racial types was rather unclear. The origins of type B were located in India because of a marked increase in type B at this location. By contrast, because all Europeans showed relatively higher incidences of type A, its place or origin could not be precisely located; they assumed it had first been introduced somewhere in North or Central Europe.[101] To test their theory effectively would require expanding their analysis even further with an even larger, more diverse subject group. To accomplish this, the Hirszfelds suggested large-scale cooperation between anthropologists and serologists—ideally "on an international basis." Again, as was the case before the war, the expense and effort of arranging this made it improbable. Confirming or refuting these group characteristics would first require a more detailed examination of all of the races in question, particularly the racial groups in Northern and Cen-

---

[98] Some theorized that, when human beings first appeared on the earth, these types were present in the same proportions. For some reason—perhaps climate, for example—these gradually altered. The Hirszfelds considered this unlikely. As proof, they referred to the fact that the Russians in Siberia had the same proportion of type B as the natives of Madagascar. Similarly, the Jews from Monastir had a blood-type distribution that clearly differed from that of the surrounding Balkan population.

[99] McCarthy and Okroi, "The Original Blood Group Pioneers ... the Hirszfelds," 26.

[100] L. and H. Hirszfeld, "Serological Differences between the Blood of Different Races," 677.

[101] The Bulgarians were believed to "have mixed with the Mongols" (ibid., 679). Theoretically, then, the Bulgarians would have had more type B. The results, however, indicated that the Bulgarians' frequency of type B was no higher than neighboring Balkan peoples.

tral Europe.[102] In addition, because their findings concerned the anthropological origins of man, the Hirszfelds pointed out that follow-up studies would also be necessary on primates ("anthropoids").[103] As a result, even with its impressive numbers and intriguing implications, the research in Salonika seemed to prompt more questions than it answered. Still, the fact was that, prior to the Hirszfelds' research, no study had been able to outline any semblance of a relationship between the trait of blood type and race. For this reason, both the English medical review *Lancet* and the French journal *Anthropologie* promptly printed the article in 1919.

Their acceptance of the article was influenced by a heightened awareness of the increasing limitations and intricacies of anthropological race science. In the early twentieth century, there had been a surge in scientific articles on collected metric and physiognomic criteria, the results of which often resulted in even more confusion and uncertainty over which trait(s) were worthy of consideration. Even the seemingly simple task of measuring physical characteristics was problematic. The fact was that physical measurements were subject to interpretation as well as human error; a "round skull" to one theorist might be a "long skull" to another, and so forth. Eventually, hair and eye-color charts were designed to both improve accuracy and enable faster data collection. However, these charts were not uniform, ironically, and many anthropologists preferred not to refer to them at all. Depending on the trait being examined, the subject might be categorized into more than one racial group; for instance, an individual might have a "Nordic skull" but "Dinaric coloring." All things considered, most racial theorists would have welcomed the introduction of a simple, objective heritable trait such as blood. The theory of two separate biochemical races, as well as objective analysis of "mixing" between the two, might help in reducing (or removing altogether?) subjectivity. Furthermore, if there had originally been two "pure" races of types A and B, determining their present composition might also prove helpful in tracing the movements of prehistoric populations.

---

[102] L. and H. Hirszfeld, "Des méthodes sérologiques au problème des races," 536.
[103] They called for the immediate research of "various stocks, primitive races, and anthropoid apes."

## Jewish Physicians and Blood Science

The contributions of Jewish physicians were critical to the development of modern blood science, both in its therapeutic and anthropological applications. Clearly, the most fundamental discovery for serology was Landsteiner's discovery that there were different blood types.[104] Prior to this, research on blood had been inconsistent, unpredictable, and occasionally lethal. Ludwik Hirszfeld's work with his mentor von Dungern confirmed that the blood types were passed down from one generation to the next. Hirszfeld's later work revealed that the statistical distribution of blood groups differed along racial lines. These findings raise the question: why was there such a marked Jewish presence in blood science?

In the years before the Third Reich, Germany was internationally recognized for its prestigious medical establishment. Physicians of Jewish descent had been central contributors to establishing this reputation in all areas of medicine. Importantly, their societal involvement was not limited to medicine. Certainly, in pre-Nazi Germany and Austria, Jews were disproportionately represented in certain professions. They made up an inordinately large number of artists, writers, musicians, and scientists.[105] Although Jews constituted only about 1 percent of the total German population from the mid-1880s to 1933, the percentage of Jews enrolled in universities was much higher, fluctuating between 4.5 and 8 percent. Jewish representation in medicine was especially conspicuous. From 1891 to 1911 Jews accounted for a "startlingly high range"— between 10 and 16 percent—of those enrolled in medical faculties.[106] A large proportion of these Jews, though, were not German Jews.

The pogroms and increasing anti-Semitic restrictions throughout Eastern Europe and Russia created a near-constant influx of Jewish refugees into the relatively tolerant nations of Western Europe. Among those fleeing were victims of Russia's *numerus clausus*, which severely limited Jewish access to university education. Some 2.75 million Jews left Russia between 1881 and 1914; the majority went west and through Germany.[107] Many of

---

[104] Gottlieb, "Karl Landsteiner, the Melancholy Genius," 21.
[105] Efron, *Defenders of the Race*, 156.
[106] Ibid., 30.
[107] Ibid., 9.

these refugees decided to stay in Germany, which swelled Jewish representation in German medicine. On the eve of World War I, Jewish Russian students represented 18 percent of the students enrolled in the faculty of medicine in Berlin, 26 percent in Leipzig, and 32 percent in Königsberg.[108] As a Polish Jew, Ludwik Hirszfeld would have been part of this foreign Jewish community receiving schooling in Germany. After attending high school in his hometown of Lodz, Poland, Hirszfeld applied to study medicine in neighboring Germany. In 1902 he entered the University of Würzburg but transferred to the University of Berlin in 1903.[109] His doctoral dissertation, "On the Agglutination of Blood" (*Über Blutagglutination*), was completed in 1907. After graduating, Hirszfeld became a junior assistant in research at the Heidelberg Institute for Experimental Cancer Research, of which Emil von Dungern was director.

Born a Jew in Austria, and therefore not part of the influx from Eastern Europe, Karl Landsteiner still would have been included in most statistics on Jews in the medical professions. He graduated from the Medical Faculty of Vienna University in 1891 and later worked in the Vienna Institutes of Hygiene and Pathology.[110] In 1908 he was appointed director of pathology and assistant professor of pathological anatomy at the Wilhelmina Hospital in Vienna.[111] Landsteiner would hold these appointments until March 1920. The experiences of Hirszfeld and Landsteiner were part of a German trend to disproportionately train and hire physicians of Jewish descent. Though it was only a small percentage of the general population, the Jewish community's strong presence in medicine made it a highly visible group within Germany. The same applied to Landsteiner's Vienna. From 1880 to 1910, because of flight from Eastern Europe, Vienna's Jewish population alone increased from 72,543 to 175,318.[112] Jews, or Jews who had

---

108 Weindling, *Health, Race and German Politics*, 9.
109 Maria A. Balińska, and William H. Schneider, *Ludwik Hirszfeld: The Story of One Life* (Rochester, NY: University of Rochester Press, 2010), 6.
110 Gottlieb, "Karl Landsteiner, the Melancholy Genius," 21.
111 Ibid., 19.
112 Or "8.6 percent" of Vienna's population: Efron, *Defenders of the Race*, 156. "At the beginning of World War I, there were approximately 90,000 Jewish foreigners in the empire. Then, during the war, about 30,000 more Jews from the territories occupied by Germany were either recruited or forced to work in Germany while additional Jewish war refugees crossed the border." Steven M. Lowenstein, Michael A. Meyer, Michael Brenner, and Paul R. Mendes-Flohr, *German-Jewish History in Modern Times: Integration in Dispute, 1871–1918* (New York: Columbia University Press, 1998), 20–21.

converted to Christianity, made up well over half of Vienna's doctors from the 1880s right up until 1938.[113]

The ongoing influx of Eastern European Jews coincided with an "explosion of higher learning" that took place in Germany between 1870 and 1914.[114] Between these years, enrollments at universities and technical schools expanded from 14,000 to 61,000 students.[115] The presence of Eastern European Jews drew attention to the existing prominence of Jews in the universities. As a result, both increasingly faced harsher anti-Semitism within German academia, as well as society in general. Despite the progress made with assimilation, the rise in extremist groups and racist propaganda, as well as incidents such as the Dreyfus Affair, indicate that anti-Semitism had a place in Western Europe as well. Even with its reception of oppressed Jews, and a certain tolerance for them, Germany was also host to a particularly virulent form of anti-Semitism. So widely publicized was the ostracism and anti-Semitism encountered by Jewish doctors in fin-de-siècle Germany and Austria that the topic was the central theme of *Der Bernhardi*, a popular play of the time.[116] In spite of the significant number of Jewish doctors in Germany in 1900—they accounted for 16 percent of all physicians—they were still often denied prestigious academic appointments, were largely excluded from research positions in theoretical medicine, and were frequently charged with malpractice.[117] The nature and perceived threat of this anti-Semitism is evident in the 1893 decision of a group of German Jews to establish a self-defense organization, the Central Association for German Citizens of Jewish Faith (*Centralverein deutscher Staatsbürger jüdischen Glaubens*).

Under these circumstances, it was not uncommon for Jewish individuals, particularly those seeking professional positions, to convert to the prevailing state religion.[118] At the time he discovered the blood types, Land-

---

[113] Steven Beller, *Vienna and the Jews, 1867–1938: A Cultural History* (Cambridge: Cambridge University Press, 1991), 38.

[114] Efron, *Defenders of the Race*, 93.

[115] Ibid.

[116] "So widely publicized was the ostracism and anti-Semitism encountered by Jewish doctors in fin-de-siècle Germany and Austria that the topic was the central theme of the Viennese Jewish playwright and physician Arthur Schnitzler's *Der Bernhardi* (1912)." Efron, *Defenders of the Race*, 31.

[117] Ibid.

[118] Stephen M. Lowenstein, "Jewish Intermarriage and Conversion in Germany and Austria," *Modern Judaism* 25, no. 1 (2005): 23–61.

steiner was Roman Catholic, the majority faith in Austria. However, he had been born Jewish in Vienna in 1868—his father, Leopold Landsteiner, was a practicing Jew.[119] Ludwik Hirszfeld, converted to Roman Catholicism before World War I. We cannot be certain of their reasons for converting, though it would be safe to speculate that anti-Semitism and career opportunism might have played a role. Landsteiner's relationship with his wife Helene Wlasto, however, suggests that the Catholic faith specifically, and not just Christianity, was not a trivial matter. In 1918, after two years of marriage, Helen finally acquiesced to her husband's wishes and severed her association with the Greek Orthodox Church to become Catholic.[120] Conversion did not necessarily remove the "stigma" of being Jewish, however, as anti-Semites frequently interpreted the act to be a kind of "inauthentic mimicry."[121]

Anti-Semitism can partly explain the disproportionate role of Jews in the development of serology. The clinical fields within medicine, into which serology would have been classified, tended to have higher concentrations of Jews compared to academic medicine. This was for two reasons. First, after 1890, clinical medicine underwent a general expansion that created room for professional advancement, and Jews flocked to the various specialties of internal medicine, gynecology, ophthalmology, dermatology, and psychiatry.[122] Second, specializing in one area of clinical medicine gave the Jewish doctor the opportunity to enter into private practice, whereas research medicine could only lead to a university position.[123] When Jews, especially unconverted ones, did accept such positions, their status in academe was usually limited. The theoretical fields were the older ones and, with a more entrenched and conservative faculty, were therefore more inclined to deny access and professional advancement to individuals of Jewish descent.[124] Serology was among the various medical specialties that developed late, after 1890. Its newness would have made it more accessible than the more established fields and

---

[119] Monika Löscher, "Eugenics and Catholicism in Interwar Austria," in Turda and Weindling, eds, *Blood and Homeland*, 301. See also Monika Löscher, *"... der gesunden Vernunft nicht zuwider ...?" Katholische Eugenik in Österreich vor 1938* (Innsbruck: Studienverlag, 2009).

[120] Gottlieb, "Karl Landsteiner, the Melancholy Genius," 19.

[121] Hutton, *Race and the Third Reich*, 111.

[122] Efron, *Defenders of the Race*, 30.

[123] Ibid.

[124] Ibid.

therefore more attractive to Jews or those of "Jewish descent." This likely helped in shaping Landsteiner and Hirszfeld's career paths.

## POSTWAR BLOOD SCIENCE

Ludwik Hirszfeld's eventual contribution to blood science through his field-work in Salonika would mark the beginning of seroanthropology, an off-shoot of existing blood science committed to determining the racial value of the blood types, as well as their worth for anthropological studies of race and racial history. Ironically, given the presence of anti-Semitism in Germany and its adherents' emphasis upon the importance of race, Hirszfeld was responsible for introducing yet another potential means of racial differentiation. The prospect was greatly welcomed by biased race theorists.

In spite of their extensive contributions, both Landsteiner and Hirszfeld would eventually leave the German medical community. Before the war, in 1911, Hirszfeld left Heidelberg to accept an assistantship at the Hygiene Institute of the University of Zurich.[125] Although he had formed a close friendship with von Dungern, over time Hirszfeld felt his work in Heidelberg to be too limited. By contrast, Zurich presented Hirszfeld with the opportunity to research the specific areas of interest previously suggested in his dissertation topic—hygiene and microbiology. In Switzerland, Hirszfeld was made an academic lecturer based on his studies of agglutination and was also named associate professor. Despite his productivity and contributions while in Vienna, Karl Landsteiner was placed on "permanent retirement" by the Austrian government in May 1920. The reasons for this decision are unclear, but it may have been motivated by anti-Semitism.[126] A letter in November of that year, from the Provincial Government of Lower Austria, informed Landsteiner of the pension to which he was entitled—an amount that has been referred to as "insufficient to provide a family with the bare necessities of existence."[127] Certainly, his work had been indispensable to the development of practical blood science. By all accounts, Landsteiner was an asset to both his employer and the Austrian government.

---

[125] Hanna Hirszfeld became an assistant at the Zurich Children's Clinic.

[126] SeeHans Peter Schwarz and Friedrich Dorner, "Karl Landsteiner and his Major Contributions to Haematology," *British Journal of Haematology* 121, no. 4 (2003): 556–565.

[127] Paul Speiser and Ferdinand G. Smekal, *Karl Landsteiner: The Discoverer of the Blood Groups and a Pioneer in the Field of Immunology; Biography of a Nobel Prize Winner of the Vienna Medical School* (Vienna: Hollinek, 1975), 59.

In the years following his discovery of the blood types, Landsteiner would receive various awards. In 1909 the Medical Faculty of Vienna University proposed his appointment to associate professor of anatomy; the "Imperial and Royal Apostolic Majesty," Franz Joseph I, was "graciously pleased" to appoint Landsteiner to this position in 1911.[128] From July 1910 Landsteiner held the position of senior medical officer (*Oberarzt*). A year after the outbreak of World War I, he was promoted to the Administrative Council (*Regierungsrat*).[129] While in Vienna, Landsteiner was very productive in his job as a coroner. In ten years he performed 3,639 autopsies, accounting for one-fifth of the institute's total workload, and wrote at least seventy-five scientific papers, in addition to teaching classes and giving instruction in pathological anatomy.[130] In spite of these accomplishments, the University of Vienna abruptly dismissed Landsteiner in 1920. Later that year he took an appointment at a medical facility in Holland. Like Hirszfeld, he felt constrained by the post's limited research potential and sought employment elsewhere. With the endorsement of the Dutch authorities and Simon Flexner, the director of the prestigious Rockefeller Institute, Landsteiner was offered a position at the Rockefeller Institute in New York City.[131]

Because of the obvious relevance of the Hirszfelds' research to racial anthropology, one would expect it to elicit an immediate response from German physicians concerned about the "specter" of national eugenic and racial decline. This was particularly the case at war's end. German eugenicists were appalled by the nation's wartime losses; of the pre-1914 population of approximately sixty-five million, there were approximately two million military dead, and half a million civilian fatalities. These losses were exacerbated by the fact that the German birthrate had been considered alarmingly low even before the war; between 1906 and 1910, 31.6 children were born for every 1,000 members of the total population.[132] These levels would continue to decline during the Weimar Republic.[133] The loss of

---

[128] This appointment did not come with an increase in salary. Ibid.

[129] Gottlieb, "Karl Landsteiner, the Melancholy Genius," 19.

[130] Starr, *Blood,* 39.

[131] Ibid., 22.

[132] D.V. Glass, *Population Policies and Movements in Europe* (Oxford University Press, 1940), 269–270. See also the table "The Prussian Survey of the Declining Birth Rate of 1912: Summary of Provincial Reports," in Weindling, *Health, Race and German Politics,* 269–280.

[133] "By 1933 Germany's birthrate was the lowest in all of Europe (except for Austria). In 1932 only 15.2 children were born for every 1,000 Germans." Glass, *Population Policies,* 269–270.

men during the war was partly to blame for this decrease, but other factors were also believed responsible. Many researchers cited the increase in sexually transmitted diseases. Syphilis, which could cause stillbirths, premature death, or birth defects, was extremely common. For eugenicists, the threat of venereal disease was made worse by "racial mixing, the ineffectiveness of natural selection, and birth control."[134] To this list, some added sexual promiscuity, youth delinquency, crime, "trashy literature," rising divorce rates, and illegitimate births.[135] Both eugenicists and race theorists feared the degeneration (*Entartung*) of the "Aryan" people. Physicians joined together to pressure the government to pass measures to deal effectively with this possibility. In postwar Germany, clinics rapidly emerged to address child and maternal health, venereal disease, alcoholism, and tuberculosis.[136] In order to keep tabs on their "predicament," and in the interest of improving the genetic plight of the "Aryan" people, screenings and surveys for various physical and psychological pathologies became standard. In fact, plans for racial biological surveys originated in the immediate postwar years, 1919–1924, and drew on advances in research on hereditary diseases and serology.[137] As a hereditary trait, blood type had both eugenic and racial implications that were applicable to prevailing concerns in the German medical community at this time. The outcome of the blood type surveys fit the postwar German agenda perfectly.

The initial publication of the Hirszfelds' study in English and French would slightly delay its reception in Germany. The first studies to follow their example were in the United States, Denmark, South Africa, and China.[138] Seroanthropology was first introduced to Germany mainly through a postwar study conducted by Hungarian researchers but published in German. This study quickly drew the attention of politically biased race scientists.

---

[134] Katja Geisenhainer, *"Rasse ist Schicksal": Otto Reche (1879–1966), ein Leben als Anthropologe und Völkerkunder* (Leipzig: Evangelische Verlagsanstalt, 2002), 120. See also Lutz Sauerteig, *Krankheit, Sexualität, Gesellschaft. Geschlechtskrankheiten und Gesundheitspolitik in Deutschland im 19. und frühen 20. Jahrhundert* (Stuttgart: Franz Steiner Verlag, 1999).

[135] Nikolaus Wachsmann, "Between Reform and Repression: Imprisonment in Weimar Germany," *Historical Journal* 45, no. 2 (2002): 415.

[136] Dorothy Porter, *Health, Civilization, and the State: A History of Public Health from Ancient to Modern Times* (London: Routledge, 1999), 200.

[137] Weindling, *Health, Race and German Politics*, 464.

[138] Schneider, "Initial Discovery," 288–289.

# CHAPTER II

# SEROANTHROPOLOGY IN THE EARLY 1920s: BLOOD, RACE, AND EUGENICS

The Hirszfelds' research in Salonika was a remarkable anthropological study for its time. Not only did it propose a new method of racially screening populations, but it had surveyed an especially large subject group (8,000 people), which was crucial if the work was to receive a positive reception in the medical community. In addition, the implications of the findings were attractive in their simple claim that there had originally been two racial types: A in the West and B in the East. Furthermore, objective comparison was made possible by the biochemical race index, an equation that classified a group studied as one of three different serological racial types, the two extremes of "Western European" and "Asiatic-African," in between which fell "intermediate," or Mediterranean peoples. In spite of this seeming transparency, the study prompted many questions. The Hirszfelds referred to the "purity" of the different races in relation to "blood mixing." Did this mean that "pure" races still existed and that they could be identified by examining blood? Could blood determine the extent of miscegenation and with whom it had occurred? If so, was it then possible to indicate when and where this "mixing" had taken place? If blood types A and B were related to "Western" and "Eastern" races, respectively, was it then the case that individuals with a certain type of blood exhibited mental or physical racial characteristics traditionally associated with that race? Reactions to the study varied widely… and often by nationality.

## Frigyes Verzár and Oszkár Weszeczky:
## Seroanthropological Research in Hungary

The study generally regarded as the first to duplicate the Hirszfelds' was published by Drs. Frigyes Verzár and Oszkár Weszeczky of the University of Debrecen (Hungary) in 1921.[139] The authors set out to establish whether different racial groups that had coexisted in and around Budapest would show variations in their blood-type distributions. Their study presented an important divergence from the research in Salonika; the Hirszfelds' subjects had lived together under the same circumstances for only about two years, whereas the groups in Hungary could trace their lineage in the area back hundreds of years. The results of an analysis of peoples who had lived together for many generations would be decisive as, in spite of evidence to the contrary, there were physicians who still theorized that blood type could change with environment. If no differences were found between the groups in Hungary, it would undermine Hirszfeld's hypothesis altogether.

The three different racial types studied by Verzár and Weszeczky included "native Hungarians" from Debrecen, Germans from several villages near Budapest, and "Gypsies" (*Zigeunern*, hereafter referred to as Roma-Sinti).[140] Centuries of German migration and settlement had created a "checkerboard of German speech islands" throughout Central and Eastern Europe.[141] The Germans examined by Verzár and Weszeczky were the descendants of Habsburg colonization efforts in the wake of the reconquest of Hungary from the Ottoman Turks at the end of the seventeenth and early eighteenth centuries.[142] Large swaths of Hungary were left depopulated by the conflict,

---

[139] In 1919, unaware of the Hirszfelds' simultaneous research in Salonika, Weszeczky researched the blood types of 457 Hungarians, eighty-one Romanians, and twelve Germans in Hungary. Despite the limited size of the test group, it was clear that the blood type distributions in Hungary differed from those in Germany and the United States. Weszeczky theorized that these variations might have stemmed from racial differences, though he recognized that they could be due to factors such as diet or climate. Determining this, however, would require more extensive research, which was prevented by enemy occupation during the war. See F. Verzár and O. Weszeczky, "Rassenbiologische Untersuchungen mittels Isohämagglutininen," *Biochemisches Zeitschrift* 126, nos. 1–4 (1921): 33–39.

[140] The subjects included 1,500 Hungarians, 476 Germans, and 385 "Gypsies." "'Sinti' refers to the Sindh River in India, and 'Roma' to 'human beings' in the Romani language, though they were generally called 'Gypsies' (from Egyptians) at the time." Deborah Dwork, *History of the Holocaust* (New York: W.W. Norton, 2003), 91.

[141] Charles W. Ingrao and Franz A.J. Szabo, eds. *The Germans and the East* (West Lafayette, IN: Purdue University Press, 2008), 5.

[142] Ibid., 4.

and successive Habsburg monarchs pursued an aggressive program of resettling the areas, sometimes with South Slav refugees but more commonly with German colonists.[143]

According to their biochemical indices, the Hungarians, Germans, and Roma-Sinti were classified as "Intermediate," "Western European," and "Asiatic African," respectively.[144] These were promising results that agreed with the categorization of the same types in the Hirszfeld study. Importantly, despite centuries of living together in the same environment, the three groups had retained their serological identities. To determine why the Hungarians had more type B blood than the German settlers, Verzár and Weszeczky reconstructed the possible historical movements of the peoples they believed responsible for the introduction of "Eastern blood" into Hungary. The Hirszfelds had referred to mixing between East and West, but only loosely, and had not specified when, or with exactly which groups, this had taken place. In their depiction of events, the lower biochemical index of the peoples of Central Europe, in the "transition zone" between East and West, stemmed from miscegenation with groups farther east.

Verzár and Weszeczky theorized that this "mixing" between the original Hungarian peoples and Easterners first took place around the seventh century. After 600 AD, they believed, the groups native to Hungary merged with groups from present-day Turkey and Bulgaria. Miscegenation between them continued throughout parts of southern Russia, north of the Crimea. After some point in time, which the authors fail to specify, the two groups separated; one remained along the Black Sea, while the other traveled both north and south. The most extensive "blood mixing" was believed to have occurred with Turkish peoples, and the examiners reported that this was most pronounced in the areas north of the Tisza River in Hungary, as a result of the settlement of the Cumans (a nomadic Turkic tribe) there in the thirteenth century. [145] Further Turkish invasion and conquest followed throughout the sixteenth and seventeenth centuries—all of which increased modern Hungarians' levels of type B blood.[146]

---

[143] Ibid., 5.

[144] The Hungarians' index was 1.6, the Germans' was 2.9, and the Roma-Sinti index was 0.6. See Verzár and Weszeczky, "Rassenbiologische Untersuchungen mittels Isohämagglutininen," 34.

[145] The Tisza River, one of the main rivers of Central Europe, originates in Ukraine, flows along the Romanian border and through Hungary and eventually into the Danube in Serbia.

[146] Verzár and Weszeczky, "Rassenbiologische Untersuchungen mittels Isohämagglutininen," 36.

The Germans examined were descended from colonists who had settled in Hungary in 1710—long after the Turkish immigrations mentioned by the authors. The Germans' higher incidence of type A blood was attributed partly to their racial origins farther west, more removed from the influence of Eastern peoples, and also to the fact that they had generally "kept to themselves" after settling in Hungary, which was evident in their marriages as well as in their retention of German language and culture. This was not consistently the case, however, as indicated by the authors' statement that, when selecting subjects, they had been careful to exclude the few who were not "entirely German."[147] The Germans' blood type distribution was essentially identical to that of German farmers in Heidelberg, the area from which the original settlers had migrated.[148]

Verzár and Weszeczky took the precaution of including only those Roma-Sinti most likely to be "racially pure"—specifically, the "wandering," or nomadic, type. Other groups, referred to as "half-settled Gypsies" (*halb angesiedelte Zigeunern*) or "mixed breeds" (*Mischlinge*), were avoided as they did not have the "pure blood" deemed necessary for the study. Roma-Sinti employed as "musicians or horse-dealers" were also excluded—presumably on the grounds that these occupations involved a higher level of interaction (and therefore miscegenation) with the local population. By contrast, the ostracized wandering Roma-Sinti, not unlike the settler Germans, had not assimilated with local Hungarians. Verzár and Weszeczky declared that the three different types of Roma-Sinti had remained "mutually isolated" (*gegenseitig isoliert*) from one another, and animosity between the groups was not uncommon.[149]

The Roma-Sinti had the highest proportion of type B blood among the groups studied, which was interpreted to be an indicator of strong Eastern origins. Based on the common belief that the Roma-Sinti first appeared in

---

[147] Ibid. The native German subjects pooled were all schoolchildren, older than ten years, to ensure that the findings would be as accurate as if they been tested on adults. Although blood type was fully developed by the first year of life, there was a common, long-held misconception that this process actually took years. The reader can only assume, based on the text, that these subjects were excluded based on the presence of "non-German" characteristics—in this case, probably their language or culture, but possibly also their appearance (or possibly a combination thereof).

[148] The biochemical index from Heidelberg was based on von Dungern and Hirszfeld's "inheritance study" of 1910. The Germans in Hungary had an index of 2.9, those in Heidelberg 2.8. See E. von Dungern and L. Hirszfeld, "Über Nachweis und Vererbung biochemischer Strukturen," *Zeitschrift für Immunitätsforschung* 4 (1910), 531–546.

[149] Verzár and Weszeczky, "Rassenbiologische Untersuchungen mittels Isohämagglutininen," 37.

Germany in 1400, Verzár and Weszeczky estimated that the ethnic group probably first came to Hungary some 500 to 600 years earlier. Verzár and Weszeczky pointed to studies by the German linguist August Friedrich Pott, who suggested the Roma-Sinti to be of the "Indian race" (*indische Rasse*).[150] The Roma-Sinti language, rooted in the popular dialect of Northern India, was related to Sanskrit. Their physiognomy, too, seemed to further confirm this; even after centuries of separation from the environment of India, the authors observed that the Roma-Sinti had "hardly changed in their physical characteristics."[151] These findings neatly supported Hirszfeld's hypothesis. The blood type distributions of the Indians, who inhabited the supposed "cradle" of type B blood, were nearly identical to those of the Roma-Sinti in Hungary. These findings were especially impressive, as they indicated that the Roma-Sinti's blood type distribution had not changed for at least 1,200 years.

Verzár and Weszeczky's study would elicit further research of blood and race. Their findings supported Hirszfeld's theory that the predominant blood type shifted from A to B as one went east across the Eurasian continent. Furthermore, the similarity between Roma-Sinti and the peoples of their ancestral home, India, confirmed that blood type did not alter with time or environment. In fact, it appeared that blood type remained a racial trait for hundreds of years, making it possible to trace a population's racial origins through blood type analysis. The Hirszfelds' study was also the first to draw a comparative serological study of Germans' blood types. For obvious reasons, while in Salonika the Hirszfelds had not been able to collect blood samples from the German forces detaining them. Germans were included in their final calculations, but these statistics were drawn from prewar studies of Germans directed by von Dungern and Hirszfeld in Heidelberg and Landsteiner in Vienna. Verzár and Weszeczky's German subject group was not only much larger, but was also isolated in Eastern Europe. Their results would more effectively prove that race was "in the

---

[150] Verzár and Weszeczky are referring to August Friedrich Pott (1802–1887), a German linguist who helped to establish the study of Indo-European etymology. See Winfred Philipp Lehmann, *Historical Linguistics: An Introduction* (London: Routledge, 1992), 29. Based on semantic (historical-linguistic) research, Verzár and Weszeczky pointed out that the Hungarians were believed to belong to the Finnish-Ugric language family—part of the Ural-Altaic language group. Turkish was reasoned to be of the same descent. As Verzár and Weszeczky noted, the ancestors of the Hungarian people were widely believed to have settled in what is now Hungary in 896 AD.

[151] Verzár and Weszeczky, "Rassenbiologische Untersuchungen mittels Isohämagglutininen," 37.

blood." And in the case of the Germans, their blood was definitely "Western European" in type— Verzár and Weszeczky repeatedly emphasized the Germans' higher levels of blood type A. The surrounding peoples, by contrast, did not have as much because they had "mixed" with Eastern types.[152] Though the Germans had lived among non-Germans for centuries, the Germans' blood revealed their "complete [racial] difference" from the natives.[153]

This finding was very relevant to the *völkisch* political movement of pan-Germanism, which aimed to expand German territories and unite all the German-speaking peoples of Europe. Both the experience of wartime occupation in the East and the catastrophic outcome of World War I for Germany and Austria-Hungary had helped to intensify popular interest in the Weimar Republic for Germans living outside of Germany.[154] Even before the war, nearly a quarter of all German-speakers in Europe lived outside the boundaries of the Reich.[155] Germans living within the post-1871 German Reich were traditionally called *Reichsdeutsche*, while those beyond these borders were deemed not of the Reich, but of the *Volk* (people), and thus called *Volksdeutsche*.[156] Germans in Germany tended increasingly to characterize such communities as "lost diasporas" and lamented their separation from the Reich along with the losses after the war.[157] To those with pan-Germanist sympathies, Verzár and Weszeczky's finding that the blood of *Volksdeutsche* in Hungary "clearly differed" from the surrounding non-German populace had obvious political implications. Even though this study had been of a more recent group of German settlers, waves of German colonization reached back into the medieval period. Perhaps blood science could be used to racially identify *Volksdeutsche* and justify their reintegration into one German state—never mind the fact that self-identification as "German" may not have held much significance for these groups. For pan-German nationalists, there was no debate about their identity. They simply wanted to reclaim "German blood" for Germany.

[152] Ibid., 36.

[153] Ibid.

[154] Pieter Judson, "When is a Diaspora Not a Diaspora? Rethinking Nation-Centered Narratives about Germans in Habsburg East Central Europe," in Krista O'Donnell, Nancy Ruth Reagin, and Renate Bridenthal, eds., *The Heimat Abroad: The Boundaries of Germanness* (Ann Arbor, MI: University of Michigan Press, 2005), 221.

[155] Ingrao and Szabo, eds. *The Germans and the East*, 1.

[156] Ibid., 3.

[157] O'Donnell et al., *The Heimat Abroad: The Boundaries of Germanness*, 219.

Less Eastern type B blood among the Germans in Hungary suggested that the Eastern influx had mainly stopped short of Germany. It also objectively demonstrated that miscegenation did affect racial purity. *Völkisch* race scientists had long warned of the perils of race mixing, and Verzár and Weszeczky indicated that it could alter a group's racial makeup. Those peoples who had not mixed retained their purity, which was evident not only among the German settlers, but also among the Roma-Sinti.

Inspired by the findings in Hungary, German naval physician Paul Steffan promptly tabulated the blood-type distributions of 500 German sailors (conveniently, their blood types were already on file).[158] He, too, noted an elevated incidence (43 percent) of type A blood. Based on this, Steffan confirmed that the blood types could be used as racial characteristics and were key to the "reconstruction" of human racial history.[159]

## Surveying "Native Germans"

With the objective of identifying "racially pure" Germans, physician Paul Steffan continued seroanthropological research. He was particularly wary of the racial status of populations in large cities or nomadic groups. Steffan intended only to examine "relatively pure races" (*verhältnismässig reinen Rassen*), and thus he resolved to strictly avoid research in urban areas, as they were more likely to have had an influx of non-Germans seeking opportunity or refuge.[160] For the most part, such immigrants were Eastern Europeans and Jews. Jews were, in large part, concentrated in urban areas; by 1914, one-quarter of Prussian Jews were concentrated in Berlin.[161] Rural Germany, generally less tolerant and offering fewer prospects than urban areas, generally attracted fewer of those whom Steffan would have categorized as "racial others." Race theorists were cautioned to be very particular in selecting a group for study because of itinerant travel throughout Germany, whether by "nomads" or others. In addition, before selecting a group

---

[158]  Steffan did not actually collect the blood samples; he simply referred to their Wassermann Tests, which were mandatory in the military during the Weimar Republic. "The 'Wassermann Test,' introduced in 1905 by German Jewish physician August von Wassermann, was a diagnostic test for syphilis based on analysis of blood serum." Weindling, *Health, Race and German Politics*, 169.

[159]  Geisenhainer, *Rasse ist Schicksal*, 128.

[160]  Paul Steffan, "Weitere Ergebnisse der Rassenforschung mittels serologischer Methoden," *Archiv für Schiffs- und Tropenhygiene* 29 (1925): 370.

[161]  Weindling, *Health, Race and German Politics*, 58.

for study, researchers had to consider which events, if any, might have made a group more susceptible to past incidences of miscegenation, and which groups might have been involved. Steffan carefully considered these matters when planning his research.

Steffan chose isolated areas where the population was less likely to have been exposed to miscegenation. He was especially interested in Schleswig, a remote northern region of Germany, as he believed it to have been "outside the reach of Slavic influence."[162] The village of Peterstal, deep in the Black Forest, was also chosen. More than half of the subjects in each group had type A blood. To Steffan this suggested that the preliminary selection process was successful and confirmed the theory that the German peoples had originated somewhere in northwest Europe. Low percentages of type B blood indicated that racial mixing with Eastern races had either not occurred or been negligible, not affecting the Germans' racial index. Steffan found this encouraging but recognized that further results were required to verify his theories.

Steffan also proposed surveying the people of North Frisia (Schleswig-Holstein), because he believed that they might represent the last remnants of original, "pure" Germans (*Urbevölkerung*).[163] The island of Pellworm, in the North Sea, would have been ideal for his purposes; however, researchers were able to collect only forty-six blood samples.[164] The same was also the case in two other areas in which Steffan expressed interest: a small fishing town, Massholm, situated on the Baltic Sea island of Öhe, and Berchtesgaden (in Bavaria). But blood samples were collected from only seventy-eight individuals in Massholm and 100 in Berchtesgaden.[165] Steffan further theorized that these "original peoples" likely still existed among the "West Frisians, Jutes, and blond Scandinavians," but, again, this theory was frustrated, as his previous efforts had been, by too-small subject groups.

---

[162] Of 502 subjects examined in Peterstal, 56 percent were found to have type A blood and only 7.8 percent type B, which gave the group a much-elevated biochemical index of 4.9. The 253 Germans in Angeln had 50.6 percent type A blood, and 7.5 percent type B, which increased their figure even further to 5.3. See Steffan, "Weitere Ergebnisse der Rassenforschung mittels serologischer Methoden," 370.

[163] Ibid., 370–371. We can assume that Steffan is referring here to the "original" (native) Germans.

[164] Dr. Spangenberg in Pellworm examined 46 subjects; 34.8 percent had type A blood, and 13 percent had type B. Under "notations," Steffan commented that the number (of those examined) was "too small" ("*Zahl zu klein*"). See Steffan, "Weitere Ergebnisse der Rassenforschung mittels serologischer Methoden," 371.

[165] Ibid., 372. These islands were sparsely populated, so the subject groups would have been small regardless.

He could only speculate, as the number examined could provide "no further insight" into the matter of the inhabitants' racial makeup.

Aside from these preferences, many of the blood type surveys of Germans had confirmed higher percentages of type A blood, enough to categorize them as "Western European." At a glance, the studies seemed consistent, but even early opinions varied widely concerning the usefulness of the science. In an article published in 1923, on the heels of the first series of blood-type surveys of native Germans, Hungarian physician Aladár Beznák, a colleague of Frigyes Verzár, pointed out that "it was still not possible to absolutely differentiate between the bloods of various people, families, or racial groups."[166] He reiterated that one could not determine from an individual's blood whether they belonged to "this or that race." Only the frequencies of different blood types could demonstrate noticeable differences.[167] Beznák recognized that type A blood was known to be more common in Germany and England, but remarked that this "strange" (*merkwürdigen*) pattern had yet to be clearly explained.[168] Plus, he found it curious that both types A and B had been found in every racial type examined. Therefore, although Beznák found Verzár and Weszeczky's research "interesting," it was not conclusive. These were valid misgivings, but the fact remained that there were certain patterns. Furthermore, seroanthropology was a new science, and many felt further research was required before abandoning it.

Wilhelm Sucker, a physician who seemed to share Steffan's racial biases, praised his colleague for having coordinated the only sufficiently extensive serological study of Germans.[169] In his own survey of 1,000 German subjects in Leipzig, Sucker found their blood type distributions to diverge slightly from those reported by Steffan. Specifically, Steffan's groups had more type A blood. Due to Leipzig's location farther east than the regions chosen by Steffan, Sucker expressed concern that his subject group might

---

[166] A. Beznák, "Unterscheidung von Menschenrassen durch Blutuntersuchung," *Kosmos* 20 (1923): 92–94. See also Imre Zs.-Nagy, "Fritz Verzár Was Born 120 Years Ago: A Personal Account," *Archives of Gerontology and Geriatrics* 43, no. 1 (2006): 1–11.

[167] Beznák, "Unterscheidung von Menschenrassen.," 94.

[168] Ibid., 93.

[169] The Leipzig subjects had 41.5 percent type A blood, and 16.5 percent B. See Wilhelm Sucker, "Die Isohämagglutinine des menschlichen Blutes und ihre rassenbiologische Bedeutung," *Zeitschrift für Hygiene und Infectionskrankheiten* 102, 3/4(1924): 482–492. See also comparative chart between Steffan and Sucker in ibid., 486.

not have been composed of racially pure Germans. Perhaps, he theorized, his study had been distorted by the presence in Leipzig of many who had either come from the East, "appeared" to be of Jewish descent, or were verifiably Jewish. To eliminate this possibility, Sucker closely scrutinized his subjects to ensure their "Germanness." He recalculated the statistics after excluding individuals with foreign or "Jewish-sounding" names. Despite the removal of potential Eastern influences, the frequency of type A remained essentially the same (it actually decreased by 0.5 percent), as did type B, which rose by 0.2 percent—a difference Sucker admitted was insignificant. [170] Such a slight increase in type B probably meant that the possibility of "Eastern influence" could be dismissed. Despite surnames that suggested non-German ancestry, the blood types indicated that it was minimal. Although Sucker's Germans had less type A blood, he declared that their serological indices were comparable to those recorded by Steffan, as well as Verzár and Weszeczky. In the space of several years, enough evidence emerged to suggest that there were serological consistencies across different racial types. By 1924 Ludwik Hirszfeld was confident that it "was now possible to differentiate between 'serological races.'"[171]

Blood scientists were not all in agreement, however. A rift was developing between *völkisch* German scientists and apolitical theorists. German Jewish physician Fritz Schiff, renowned for his contributions to serology, published an article with his colleague, Heinrich Ziegler, that disputed a connection between blood and race. Believing that Verzár and Weszecky's work was responsible for the recognition of blood type as "a highly sensitive means to prove or disprove differences in the descent of individual groups of people," Schiff and Ziegler decided to research the matter themselves.[172] Their subject group included 750 patients in a Berlin hospital and 230 German Jews. These figures were further supplemented by statistics previously collected by other researchers.[173] The results suggested that

[170] Blood type distributions of the 730 Germans in Leipzig: type AB 7 percent, type A 41.4 percent, type B 16.7 percent, type O 34.9 percent. Ibid., 487–488.

[171] Ludwik Hirszfeld, "Die Konstitutionslehre im Lichte serologischer Forschung," *Klinische Wochenschrift* 3, no. 26 (1924): 1180.

[172] F. Schiff and H. Ziegler, "Blutgruppenformel in der Berliner Bevölkerung," *Klinische Wochenschrift* 3, no. 24 (1924): 1078.

[173] Ibid. Instead of actually individually drawing the blood, they primarily gathered this data "indirectly," from types already recorded during Wassermann Reaction tests (for syphilis). Separate comparisons of gender and age did not appear to influence the results. See also chart on page 1078 on Germans throughout Europe.

Verzár and Weszeczky may have been mistaken in their contention that blood was a racial trait. Because of their supposedly non-Western origins, the Jewish subjects theoretically should have had higher levels of Eastern type B blood. However, Schiff and Ziegler's statistics indicated that non-Jewish Germans and German Jews were not serologically different but were, in fact, "strikingly similar" to one another.[174] The percentages were so close that one could not discern between the "blood of Jews and non-Jews." Schiff and Ziegler attributed this to the possibility of miscegenation, or to the influence of the environment on the subjects' blood— indicating that their permanence was still not recognized throughout the German medical establishment. Regardless of the source, the findings caused Schiff and Ziegler to question the existence of a relationship between blood and race. However, their doubts were largely ignored by other German blood scientists.

Interest in seroanthropology continued into the postwar years, growing more complex as researchers attempted to align the blood types with other traits examined in conventional racial analyses. Polish physicians W. Halber and Jan Mydlarski surveyed the blood types, as well as various physiognomic characteristics, of a large contingent of Polish soldiers. They recorded the blood types, skull, nose, and face shapes of 11,488 men. The results of their research were published in a German medical journal, the *Zeitschrift für Immunitätsforschung und experimentelle Therapie* (Journal of immunity research and experimental therapy).[175] Their findings suggested a link between popular racial physiognomic stereotypes and their Western and Eastern serological types. Halber and Mydlarski reported that individuals with type A blood tended to have "longer skulls, smaller noses, and narrower faces," which led them to express "no doubt" that type A blood corresponded with Nordicism. By contrast, because of their rounder heads, medium-sized noses, and "particularly large faces," those with blood type B tended to be "Slavic" (*Laponoidal*) in type.[176] Individuals with type

---

[174] Ibid.

[175] Wanda Halber and Jan Mydlarski, "Untersuchungen über die Blutgruppen in Polen," *Zeitschrift für Immunitätsforschung und experimentelle Therapie* 43 (1925): 470–484.

[176] A total of 11,488 Polish soldiers were examined; 37.6 percent had type A blood and 20.9 percent had type B. For the distributions by individual regions, see Table 30, "Verbreitung der Blutgruppen in Polen nach Provinzen." In Hirszfeld, "Über die Konstitutionsserologie im Zusammenhang mit der Blutgruppenforschung," 431 and 441.

O blood were not easily categorized as either Nordic or Slavic, with their unpredictable mix of "long and medium skulls," smallish noses, and "very narrow faces," which led the authors to classify type O as "Mediterranean."[177] Halber and Mydlarski's research marked the beginning of a trend in the early 1920s to supplement seroanthropological studies with other traits, in the hope that this would lend some insight where blood alone could not. Because of the study's respective association between type A and B blood and Nordic and Slavic physiognomies, it was subsequently often cited by German blood scientists.

Working from the results of his own and others' research, Paul Steffan proposed in 1925 a revised version of the Hirszfelds' theory of disparate serological types. Instead of the category "Western European," Steffan labeled those races with the highest proportions of type A "Atlantic," and those groups with type B were termed "Gondwanic," in lieu of "Asiatic-African." Like Hirszfeld, Steffan maintained that the two types had separate geographic origins. Each type was at its racially purest in the area from which it had originated, or from what Steffan referred to as its "serological pole." Based on his findings of particularly high levels of type A blood there, Steffan placed the Atlantic (type A) pole in the "purely Nordic" area of Schwansen, in northern Germany.[178] (See Figure 3.) However, he noted a possibility that the A pole was much farther north—perhaps among the Swedes, who were, he mused, even more Nordic than the Germans; this was only a suggestion as, of 1925, no blood type studies had yet been conducted on the Swedes. Steffan located the "B" or "Gondwanic" pole at the opposite end of the Eurasian continent, in Peking. (See Figure 4.)

The similarities to the conclusions based on Hirszfeld's Salonika research are obvious. In both studies, the focus was solely on blood types A and B, repeating the familiar claim that A had originated in the West and B in the East. A and B were depicted as literal opposites, and Steffan maintained that movement away from these original points had resulted in "blood mixing"—

---

[177] Results summarized in René Dujarric de le Riviere and Nicolas Kossovitch, "Les groupes sanguins en anthropologie," *Annales de Médecine Legale, Criminologie, Police Scientifique et Toxicologie* 4 (1934): 275–294. Maps and a summary in German can be found in Steffan, ed., *Handbuch.*

[178] Hirszfeld, "Über die Konstitutionsserologie im Zusammenhang mit der Blutgruppenforschung," 450. According to Steffan, the core of the "Atlantic" pole included the North European and Alpine races; he considered it very possible that the Native Americans of North America belonged to this category as well. Steffan lamented the fact that the Swedes—quite possibly the "purest" (*reinsten*) Northern Europeans—had not yet been examined. See Steffan, "Weitere Ergebnisse der Rassenforschung mittels serologischer Methoden," 378.

Figure 3. Steffan's depiction of the Atlantic "type A" pole in northern Germany. Paul Steffan, "Weitere Ergebnisse der Rassenforschung mittels serologischer Methoden," *Archiv für Schiffs- und Tropenhygiene* 29 (1925): 369–391.

a phrase identical to that used by the Hirszfelds.[179] There are also, however, conspicuous differences in their tone. Steffan's analysis of the results is clearly racially biased. Steffan used "Nordic" instead of "Western European" and substituted "Gondwanic" for "Asiatic-African." "Gondwanic" was deliberately chosen to imply that these peoples were less evolved (Gondwana was a prehistoric supercontinent). Also categorized as Gondwanic were other groups viewed as "primitive" or "racially inferior" in Nordic circles, such as the Burmese, Indians, Africans, and Aborigines.[180] Furthermore, while the

---

[179]  Steffan affiliates the results with four possible areas: North Atlantic Pole, East Atlantic Mixed Area, Gondwanic Mixed Area, and Gondwanic Pole. Steffan, "Weitere Ergebnisse der Rassenforschung mittels serologischer Methoden," 378.

[180]  According to Steffan, the geological history of these areas indicated that they had at one point been connected to South Africa, Madagascar, India, and, even farther back, with the Australian continent—a prehistoric supercontinent known geologically as "Gondwana." Hirszfeld, "Über die Konstitutionsserologie im Zusammenhang mit der Blutgruppenforschung," 450. Steffan refers to the expansion of Mongolians and Negroes— particularly the "primitive" peoples such as the *Bergweddas* [central African dwarf people], *Australnegger* [Aborigines], and the *innerafrikanischen Pygmäenstämme* [the Pygmies of middle Africa]." Steffan, "Weitere Ergebnisse der Rassenforschung mittels serologischer Methoden," 379. Listed in an index table (42) are the peoples and races of less A and of increasing type B blood. Ibid., 378.

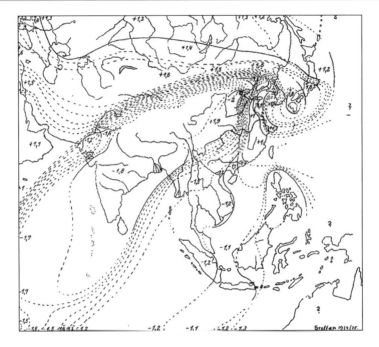

Figure 4. Steffan's depiction of the "Gondwanic type B" pole in Peking. Paul Steffan, "Weitere Ergebnisse der Rassenforschung mittels serologischer Methoden," Archiv für Schiffs- und Tropenhygiene 29 (1925): 369–391.

Hirszfelds theorized that type A had originated "somewhere" in Europe, Steffan situated the origins of Nordic type A blood specifically in Germany despite, by his own admission, a lack of complete statistics. His contention that type B originated on the farthest eastern edge of Asia, instead of in India, emphasized even further a contrast between the types.

Unlike the Hirszfelds, Steffan did not mention anywhere in his study that all four blood types had been found in every racial group examined, or that seroanthropology was based on differences in frequency, not kind. Through these omissions, Steffan implied that racial categorization via blood was a simple procedure.[181]

---

[181] The Hirszfelds had their bar graph; Steffan had his map. In 1925, Steffan presented the first worldwide "cartographic" representation of the serological types. Based on these images, Steffan made additional incredible speculations on global racial migration patterns, and even continental drift. See Steffan, "Weitere Ergebnisse," 388. Other methods were made to analyze race through arithmetical means. In 1926 Otto Streng presented a geometrical theorem in which each blood type was assigned a particular point in a triangle. By use of this methodology, Streng felt that he could demonstrate an "ethnic type" and therefore suggested that racial classification might be better represented through serology than other anthropological characteristics. Streng suggested that his methodology was more concise than other techniques, while in practice many found that

Steffan applied his theory of serological poles to Germany itself. The high frequency of type A blood in Peterstal in the Black Forest did not surprise him, Steffan explained, as these people were members of the "Alpine race"—a type that had not been exposed to Gondwanic (Eastern) blood.[182] As for the village in Schleswig, Steffan believed that it had served as a passageway for Northern Europeans headed south. He could not be certain, interestingly, because some of the population was unwilling to be tested; only 253 individuals participated. For theoretical purposes, Steffan supplemented the group with figures from Schwansen, in Schleswig-Holstein, an area whose population, he believed, had a similar racial makeup as those people in the Black Forest. It was as Steffan suspected—their "racial purity" was still intact. Previous physiognomic studies of the "blond, blue-eyed, lightly pigmented" Northern European people throughout England, France, and Germany led theorists to suspect that they had "kept free of" racial mixing. This was proved through blood-type analyses. Steffan observed that all the blood type surveys conducted in Northern Europe revealed that these peoples were members of the Atlantic (or Nordic) race.[183]

Despite his placement of the A-pole in Germany and his repeated references to Nordic peoples in various German regions, Steffan's interpretations reveal his suspicion that not all Germans were "Nordic." He wanted to examine subjects in Berchtesgaden not because of their supposed "racial purity," but because some researchers theorized that these Germans were "Dinarian"—a racial type presumed common in Eastern Europe, believed to be the result of mixing between Western and Eastern peoples. According to Steffan, Dinarians were the first Europeans to have experienced significant "Gondwanic" influence.[184] Blood type research showed that Germans in this area did tend to have more type B blood than those to the north and west. Steffan attributed this increase to earlier migrations of Eastern peoples, which he believed had been relatively common from the European foothills up to West Prussia, Posen, Berlin, Leipzig, and even in more isolated areas, such as the "Polish-settled" industrial province of Westpha-

---

his indices of serological similarity actually disguised rather than clarified such correlations. See Steffan, ed., *Handbuch*, 15.

[182]  Steffan, "Weitere Ergebnisse der Rassenforschung mittels serologischer Methoden," 370.

[183]  Ibid., 372. Steffan uses the terms "Nordic" and "Atlantic" interchangeably.

[184]  They were suspected to be related to the "alarodischen" (Caucasian) races. Ibid., 371.

lia.[185] The prospect of employment had drawn many Eastern Europeans to Westphalia; Steffan referred to their influx as "extremely strong." Steffan remarked that one only had to look at the peoples of Westphalia and their physical traits (*Stigmata*) for confirmation. Steffan was clearly preoccupied with the racial makeup of the German people. These were *völkisch* concerns: were the subjects "Nordic"? Were they still "pure"? If mixing had occurred, what was the extent of the "damage"? Steffan's wariness of urban areas surfaces again and again. Steffan complained that miscegenation in Westphalia had become "so extreme" that it was comparable to that of a large city.[186] To Steffan, the increase in type B blood among Westphalians served as further evidence of racial decline.[187]

Westphalia was not unique; other studies also reported increased levels of type B blood. This was also the finding of physicians W. Klein and H. Osthoff of the Municipal Health Office in Herne, who sampled the blood and recorded various physiognomic characteristics of 1,229 schoolchildren.[188] Like Steffan, Klein and Osthoff theorized that the "non-Nordic" incidence of type B was the result of an influx of foreign workers seeking employment, drawn to the area from the eastern provinces, but also from the Rhineland and Westphalia. To differentiate among their subjects, they classified them as either "Eastern, Western, or Mixed" (*Mischlinge*).[189] The patterns they observed suggested "Eastern" influence. As in Halber and Mydlarski's study, "Eastern blood" matched an Eastern appearance. Blood type B was found to be more common among brunets than blonds; it was also more common among round skulls as opposed to "long" ones.[190]

Others were more cautious when interpreting results. A 1926 article in *Eugenical News* disputed a connection between appearance and blood type.[191] Ella F. Grove, a researcher at the Cornell University Institute of Immunological Research, traveled to Japan and the Philippines to study

[185] Hirszfeld, "Über die Konstitutionsserologie im Zusammenhang mit der Blutgruppenforschung," 451.

[186] Steffan, "Weitere Ergebnisse der Rassenforschung mittels serologischer Methoden," 375.

[187] Dr. Klein's blood type survey of 718 individuals in the city of Herne in Westphalia, with its higher incidence of type B blood, seemed to confirm Steffan's theory.

[188] W. Klein and H. Osthoff, "Hämagglutinine, Rasse und anthropologische Merkmale," *Archiv für Rassen- und Gesellschaftsbiologie* 17, no. 4 (1926): 371.

[189] Magnus Hirschfeld, "Über die Verwendung serologischer Methoden zur Rassenforschung," *Zeitschrift für Tierzüchtung und Züchtungsbiologie* 8, no. 1 (1926): 135–136.

[190] Hirszfeld, "Über die Konstitutionsserologie im Zusammenhang mit der Blutgruppenforschung," 438–439.

[191] E.F. Grove and A.F. Coca, "On the Value of the Blood Group 'Feature' in the Study of Race Relationships," *Eugenical News* 11, no. 6 (1926): 89–91.

the Ainu peoples, a group in which the United States National Research Council expressed particular interest because they were "pure blooded" and believed to be threatened with (racial) extinction. The same principle applied when the Americans chose their subject group as when Steffan had chosen his; isolated groups of "unmixed" peoples were preferred for sero-anthropological study. The Ainu were regarded as racially distinct from surrounding types and had remained separated throughout the generations. Their physiognomy was consistent; Grove described how all the Ainu had "abundant black hair, black eyes, and the same placement of the eyes and other features, as well as the same skin color."[192] Unlike the surrounding Asian peoples, the Ainu were categorized as Caucasian.

Based on her examinations of the blood and physiognomy of two separate Ainu groups, Grove concluded that blood type was not useful in racial classification.[193] The supposed uniqueness of the Ainu racial type was not apparent in the group's blood type distribution. Instead, their blood placed these "white-skinned, hairy peoples" into the same serological category as the "'racially dissimilar' Senegal Negroes, Sumatra Chinese, Annamese Malays, and Javans."[194] Grove observed that blood tests clearly could not be used for racial differentiation if such "widely different races" shared the same patterns of type distributions.[195]

The conflict between Klein and Osthoff's conclusions on the one hand, and Grove's on the other, represents a larger debate that had already developed in seroanthropology by the mid-1920s as to whether there was a link between blood type and appearance. Many subjects who "looked Aryan" turned out to have type B blood. There were other possible connections that could be examined, however.

Race theorists believed that every race possessed its own characteristic interconnection of both physical and psychological traits.[196] Biased researchers credited the Nordic mentality with "industry, vigorous imagination, intelligence, foresight, organizing ability, artistic capacity, individu-

---

[192] Ibid., 90.

[193] The two groups studied were 442 "Sulu Moros" and 500 "Samal Moros."

[194] E.F. Grove, "On the Value of the Blood-Group Feature As a Means of Determining Racial Relationship," *Journal of Immunology* 12, no. 4 (1926): 260.

[195] Grove and Coca, "On the Value of the Blood Group 'Feature' in the Study of Race Relationships," 91.

[196] Richard T. Gray, *About Face: German Physiognomic Thought from Lavater to Auschwitz* (Detroit: Wayne State University Press, 2004), 243.

alism, willingness to obey orders, and devotion to [a specific] plan or idea."[197] To these, race theorist Fritz Lenz added the traits of "self-control, self-respect, respect for life and property, desire to know the unknown, a certain wanderlust, and a fondness for the sea."[198] Anthropologist Otto Ammon classified the "Aryan longhead" as prepared "to risk death and to work for a poor financial reward in order to attain a higher ideal."[199] Völkisch theorists believed that "Nordic individuals" had stronger constitutions than other racial types, which was supposedly evident in their lower incidence of mental and physical disorders. It made sense to expect, that they would also then be less likely to be incarcerated in asylums, convicted of a capital crime, or engaged in similarly immoral conduct. The number of "mentally or physically degenerate" "Aryans" was believed to be minor compared to the frequency and severity of such disorders among the "less noble" races. If type A individuals "looked Aryan," it would be reasonable to believe that they would be more inclined to "act Aryan" as well. As one symptom of this belief, völkisch blood scientists tended to emphasize studies that suggested a relationship between blood type and temperament—specifically type A with mental stability, and type B with mental disorders. The same reasoning applied to diseases, as Aryans were believed to have a stronger mental and physical constitution, and as such, type A persons would be accordingly less susceptible to disease or genetic afflictions. One physician summed it up best when he remarked that a relationship between blood type and illness would prove "important in light of racial studies."[200]

## BLOOD TYPE AND GENETIC INFERIORITY

Studies of blood and pathological conditions were common and were not necessarily associated with race theory—most researchers were interested in how blood type might alert physicians to predispositions to illness, or which therapies might work better for specific diseases. Examining the parties in question was certainly made easier by the fact that the subjects were often conveniently housed together in one place, such as hospi-

---

[197] According to German race theorist Eugen Fischer. See Proctor, *Racial Hygiene*, 56.
[198] Ibid.
[199] Weindling, *Health, Race and German Politics*, 167.
[200] L.H. Snyder, "Human Blood Groups: Their Inheritance and Racial Significance," *American Journal of Physical Anthropology* 9, no. 2 (1926): 250.

tal and asylum inmates, and convicts in prisons. There were findings that corresponded with *völkisch* race theory. In 1924 doctors Franz Schütz and Edgar Wöhlisch, of Kiel and Würzburg universities, respectively, found elevated levels of type A blood among college graduates and more type B among prisoners.[201] The same pattern was reported by the anthropologist Max Gundel, also at the University of Kiel. Gundel observed marked differences in blood type distributions between those with mental disorders and those without; there was a "significant increase" (*bedeutende Zunahme*) of type B among the asylum inmates examined, who had 25.1 percent— a much higher percentage than the 10–12 percent within the lay population.[202] Suspecting a link between blood type B and mental instability, Gundel proceeded to research criminals' blood types from various institutions in the region of Schleswig-Holstein.[203] While the surrounding noncriminal population had only 12.5 percent type B blood, the prisoners had a much higher 19.1 percent.[204] Gundel again referred to a disproportionate level of blood type B among the prisoner population.[205] After retabulating his results, Gundel was able to make this discrepancy even more dramatic. He reclassified the subjects into three groups based on the severity of their crime and then recalculated their blood type distribution. Group I included the blood types of all prisoners found guilty of one or more of the following crimes: murder, attempted murder, infanticide, child murder, manslaughter, attempted manslaughter, robbery, and assault. Group II consisted of those who had been convicted of theft, attempted theft, and/or

---

[201]  Franz Schütz and Edgar Wöhlisch, "Bedeutung und Wesen von Hämagglutination und Blutgruppenbildung beim Menschen," *Klinische Wochenschrift* 3, no. 36 (1924): 1614–1616.

[202]  Gundel examined 402 asylum inmates: 39.3 percent were type A, and 25.1 percent were type B (see table on page 52 in Hirszfeld, "Die Konstitutionslehre im Lichte serologischer Forschung," 1924). Ludwik Hirszfeld commented that it was "notable" that Gundel found a higher percentage of type B individuals than the average in Schleswig-Holstein. See Hirszfeld, "Über die Konstitutionsserologie im Zusammenhang mit der Blutgruppenforschung," 481.

[203]  The prisons were in the cities of Neumünster, Rendsburg, Altona, Glückstadt, and Flensburg. A total of 884 blood samples were taken. The results were published on November 12, 1926. See Max Gundel, "Rassenbiologische Untersuchungen an Strafgefangenen," *Klinische Wochenschrift* 5, no. 46 (1926): 2165. See also Max Gundel, "Einige Beobachtungen bei der Rassenbiologischen Durchforschung Schleswig-Holsteins," *Klinische Wochenschrift* 5 (1926): 1186.

[204]  Gundel, "Rassenbiologische Untersuchungen," 2165. These statistics were contrasted with the Germans in Schleswig-Holstein; of 3156 examined, 41.3 percent were type A, and 12.5 percent were type B. See Hirszfeld, "Über die Konstitutionsserologie," 430.

[205]  Gundel, "Rassenbiologische Untersuchungen," 2165. Compared to the general population of Schleswig-Holstein, the prisoners overall had considerably lower levels of type AB (this had dropped from 5 to 2 percent).

"concealing stolen property for profit." Arsonists made up the final group.[206] Group I prisoners were considered to be the most heinous offenders: 30 percent of this group had type B blood, and this gradually lessened with groups II and III to 18.2 and 3.2 percent, respectively.[207] In another graph comparing length of sentence and blood type, Gundel found type B to be much more common among individuals who received nine years to life— the harshest sentences.[208] To Gundel, these results confirmed the fact that blood type B was "much more frequent" among severe and repeat (*rückfällige*) felons.[209]

The rather hasty interpretation of his findings strongly implies Gundel's political preferences. Gundel was aware of the classification of type B blood as Eastern, or "non-Nordic." Despite the fact that he had examined less than 1,000 subjects, and all from the same geographic region, Gundel made bold claims concerning the criminal tendencies of those with type B blood. Not only was type B more widespread among felons in general, but it was especially pronounced among the worst prisoners—those found guilty of offenses such as murder, manslaughter, and rape. Many within this group had been repeat offenders, and some were so delinquent as to be categorized as "beyond reform." The pattern was reversed concerning type A blood, which was less common among the prison population overall (when compared to law-abiding Germans).

Gundel's claims that type B blood was a marker of both racial and eugenic inferiority were particularly applicable during the Weimar Republic, amidst anxieties about the cost and care of the mentally and physically incapacitated, as well as the rising crime rate. Most Weimar prison officials and criminologists were convinced that certain individuals were destined to offend repeatedly, and that biological factors played an important role in making these offenders "incorrigibles" (*unverbesserlich*).[210] While some might have been suitable for rehabilitation, authorities often believed that these "incorrigibles" had to be isolated for lengthy periods of time—in most cases, for life.

---

[206] Ibid., 2166. Gundel refers to the groups as "A, B, and C," but I have substituted Roman numerals to prevent confusion.

[207] Ibid.

[208] Blood type A distributions in the three groups: I 46.8 percent, II 42.5 percent, III 33.3 percent.

[209] Gundel, "Rassenbiologische Untersuchungen," 2166.

[210] Nikolaus Wachsmann, "Between Reform and Repression: Imprisonment in Weimar Germany," *Historical Journal* 45, no. 2 (2002): 423.

The Weimar-era shift in categorizing criminals was strongly influenced in Bavaria by the prison doctor Theodor Viernstein, a key figure in the criminal-biological movement and a fanatical racial-hygienist.[211] In several high-profile lectures in 1930, Viernstein claimed that half of all inmates were in the dreaded "incorrigible" category, and that this was mostly for hereditary reasons.[212] Gundel's study would be repeatedly cited because of its relevance to larger German social and political concerns. Other German physicians who duplicated Gundel's research on separate "eugenically unfit" subject groups reported a similar increase in type B blood.[213] Researchers W. Dölter and H. Heimann reported that it was higher among epileptic and paralyzed individuals. Numerous studies pointed to type B's higher levels among such groups as "imbeciles, manic-depressives, psychopaths, and hysterics," which led others to conclude that there was in fact a definite correlation between mental health and blood type.[214]

Racist anthropologists further anticipated more frequent diagnoses of physical illness among individuals with type B blood. Still, extensive serological studies of sick individuals took place only after World War I. In 1921 an American physician commented that analyses of blood and disease had received "practically no consideration."[215] In that same year one of the earliest surveys of blood type and disease was published by British physicians J. Arthur Buchanan and Edith T. Higley. Curious about whether there was a "fixed relationship" between the two, they recorded the blood types of numerous subjects suffering from a range of illnesses.[216] These results were inconclusive, though other such studies quickly followed during the interwar flurry of blood type surveys. No consistent pattern was recognized. In a study of 250 Russians suffering from tuberculosis, type B blood was found to be more common.[217] In a German study, however, both types A and B were elevated. One study in Moscow analyzed a group of malarial subjects

---

[211] Ibid.

[212] Ibid., 429.

[213] See Ernst Kretschmer, *Körperbau und Charakter* (Berlin: Springer, 1926).

[214] See B. Chominskij and L. Schustowa, "Zur Frage des Zusammenhanges zwischen Blutgruppe und psychischer Erkrangung," *Zeitschrift für Neurologie* 115 (1928): 304.

[215] J.A. Buchanan and E.T. Higley, "The Relationship of Blood Groups to Disease," *British Journal of Experimental Pathology* 2 (1921): 247.

[216] Including carcinoma, pernicious anemia, lymphatic leukemia, myelogenous leukemia, splenic anemia, hemophilia, purpura, kidney disease, cardiac/valvular/myocardial disease, thyroid disease, non-cancerous tumors, calculi, fibroid uterus, gall-bladder disease, ulcers, and jaundice.

[217] The increase in type B was also associated with a lower frequency of type O blood.

and, oddly, reported a predominance of type AB blood. Another English researcher reported the same, though with only 100 individuals, his subject pool was insufficient. By the mid-1920s, as with racial studies of blood, eugenic studies of blood had become a contentious topic. In 1926 Ludwik Hirszfeld believed that certain diseases could be correlated with certain blood types.[218] In that same year, however, an American serologist claimed that there was no "conclusive evidence" of a relationship between blood type and disease.[219] Many other physicians, reporting from various nations on an assortment of diseases, came to the same conclusion and were not able to detect any difference in blood type distributions between their sick and healthy subjects. German examiners Hallo and Lenard could not detect any difference between the blood type distributions of consumptive and healthy individuals. A Russian physician who examined 532 patients with scarlet fever also found their blood type frequencies to be comparable to those of the general population. Inconsistencies and contradictions would continue to plague analyses of blood and disease.

By the mid-1920s, the Hirszfelds' original theory of different serological types had been broadly received and reproduced on an international scale. In 1926 one scientist commented that biochemical indices of various nationalities had been determined "everywhere."[220] The initial appeal of the study can be attributed in part to its relatively simple methodology and its objectivity. Unlike racial physiognomy, blood typing did not involve the use of color charts to assess a subject's hair, skin, and/or eye color, or calipers for skull measurements. With little apparent ambiguity, a clumping reaction indicated which one of the four types the sample was. All of these efforts were working towards the prospect of an improved means of racial classification. Nonetheless, interest in seroanthropology was spotty. Instead, it tended to vary by country; an indication of the geographical dimension of the research can be found in the nationality of the authors.[221]

---

[218] Hirszfeld, "Über die Konstitutionsserologie," 423. Ludwik and Hanna Hirszfeld did publish a study on blood type and diphtheria; see L. and H. Hirszfeld, "Étude sur l'hérédité en rapport avec la sensibilité à la diphtherie," *Comptes Rendus des séances de la société de Biologie* 90, no. 15 (1924): 1198–1200.

[219] Snyder, "Human Blood Groups: Their Inheritance and Racial Significance," 251.

[220] See Table XI, "Percentages of the Four Blood Groups among Various Peoples, with the Frequency of Each Factor Concerned." Includes all (international) blood type surveys up to 1926. Snyder, "Human Blood Groups: Their Inheritance and Racial Significance," 245–247.

[221] William H. Schneider, "The History of Research on Blood Group Genetics: Initial Discovery and Diffusion," *History and Philosophy of the Life Sciences* 18, no. 3 (1996): 287–288.

Because of its initial publication in English and French, the Hirszfelds' research first gained the attention of physicians in Great Britain, the United States, and France. With the publication of Verzár and Weszeczky's study, interest in Central and Eastern Europe followed shortly thereafter; even Ludwik Hirszfeld referred to Verzár and Weszeczky's work as "interesting and important," noting its further confirmation that blood types could serve as "racial characteristics."[222] After several years of postwar research, another pattern emerged. Interest in seroanthropology gradually began to decline in Britain and America, while it gathered momentum throughout Central and Eastern Europe.[223]

## VÖLKISCH RESEARCH

Certain traits of racially biased blood type analysis distinguished it from impartial research. Nordic blood scientists strongly emphasized results that corresponded with their racial beliefs. Seroanthropology, in particular the Hirszfelds' landmark study of 1918, was favored because it implied significant racial differences between the peoples of the East and West, which pan-Germanists had been touting since the nineteenth century. Of course, it also associated the German people with "Western European" type A blood. For this reason, Paul Steffan, and others of his racial bias, eagerly reported elevated frequencies of type A blood among the Germans he studied, but failed to acknowledge evidence to the contrary (i.e., Schiff and Ziegler). Steffan and his colleagues also believed that there was a relationship between blood type and how one looked. To verify that the two were related, many would refer to studies indicating that "Nordic long skulls" and light pigmentation were more common among subjects with type A blood, and "round heads" and darker coloring among type B individuals. Because of its purported origins in the East, far from the "Aryan peoples," type B was considered to be an indicator of racial and eugenic inferiority. This perception also explains why certain analysts were so quick to draw correlations between type A blood and positive traits, and

---

[222] Hirszfeld, "Über die Konstitutionsserologie im Zusammenhang mit der Blutgruppenforschung," 437–438.

[223] See Marius Turda, "From Craniology to Serology: Racial Anthropology in Interwar Hungary and Romania," *Journal of the History of Behavioral Sciences* 43, no. 3 (2007): 361–377 and idem, "The Nation as Object: Race, Blood and Biopolitics in Interwar Romania," *Slavic Review* 66, no. 3 (2007), 413–441.

type B and negative ones, respectively. These tendencies were evident in the contention that type A blood was elevated among college graduates, while type B was more common among mental patients and hardened criminals.

In fact, blood type surveys revealed differences between ethnicities, but seroanthropology was hardly an exact science. In Verzár and Weszeczky's study, type A blood was 43.5 percent among the Germans but only 39.7 percent among the Hungarians. The Germans, in turn, only had 12.6 percent type B blood, while it was 18.8 percent among the "Magyar" Hungarians. What certainly appear to be negligible differences were repeatedly emphasized by the authors as distinct racial differences. Indeed, they claimed that the Germans had retained their "complete racial difference" from the surrounding Hungarians and Roma-Sinti. Based on these seemingly slight differences in percentages, Verzár and Weszeczky, Hirszfeld, and Steffan all claimed that the blood types could be used in determining race. Others implied that, on account of their disparate origins, the blood types could be used as a marker of inferiority. This was evident in Gundel's fixation on the presence of type B blood among asylum inmates and prisoners. In reality, the highest percentage of type B recorded by Gundel was 33 percent, which meant that the vast majority of the subject group did not even have type B blood. At 33 percent, the incidence of type B was much higher than the reported 13 percent type B for the general population of Schleswig-Holstein. But it was hardly enough of a difference to propose, for instance, penal system reform. Other examiners also made sweeping generalizations based on rather loose differences; Halber and Mydlarski, as well as Klein and Osthoff, claimed that their subjects with type B blood tended to have the Eastern traits of darker hair and "rounder heads." Steffan also broadly referred to the "Mongoloid" (*mongolide*) bodily characteristics of the peoples of Central Europe, characterized by elevated levels of type B blood.[224] Simply put, seroanthropology required exaggeration to give the impression that it was useful. Making such statements made it necessary to overlook potential flaws or contradictions in their theory. For instance, nowhere in their studies did Gundel, Steffan, or Verzár and Weszeczky consider the fact that each of the four blood types was present in the population—a crit-

---

[224] Steffan, "Weitere Ergebnisse der Rassenforschung mittels serologischer Methoden," 384–385.

icism commonly leveled by others. Nor did any of them present their theory with the stipulation that further research would be necessary before implicating type B blood as a marker of mental instability. Omissions were hallmarks of biased racial research—regardless of the trait (blood type or other) in question.

Seroanthropology was an unpredictable and fickle science. Comparison between studies revealed that blood type distributions could vary considerably even within the same racial types. Ironically, it shared this drawback with the physiognomic characteristics it was hoped that it would replace. For instance, while Hirszfeld reported a blood type index of 4.5 for the English in 1918, a later study by authors Buchanan and Higley—again of English subjects—yielded a much lower 2.5. Among the Italians he examined, Emilio Rizzatti calculated a value of 5.4, yet his colleague Ruggero Romanese's survey came to only 2.1. Schiff and Ziegler reported a biochemical index of 2.1 among Berliners. The findings prompted a response from German Jewish physician Magnus Hirschfeld, who campaigned for sexual reform and against anti-Semitism and certainly was attuned to attempted misappropriations of science. He pointed out that this was quite close to the Jews' 2.7—a difference so negligible that both would have been classed together as "intermediate" types, according to the Hirszfelds.[225] The mixture of results led Magnus Hirschfeld to question altogether the "usefulness of the blood types in racial studies."[226] Another examiner claimed that the "cradle" of the individual races could apparently not be learned from blood type data alone.[227]

Opinions on a possible connection between blood type and its manifestation in physiognomy were similarly diverse. Stemming from his pre-1914 research on blood types and physiognomy, Ludwik Hirszfeld strongly suspected that there was no fixed relationship between blood group and appearance.[228] In Salonika a cursory examination of some of the subjects' appearances seemed to confirm this. Nonetheless, the matter was still pursued in the hope of finding a relationship between the conventional physical characteristics so heavily relied upon and blood—

---

[225]  Hirschfeld, "Über die Verwendung serologischer Methoden zur Rassenforschung," 136.

[226]  Ibid.

[227]  Snyder, "Human Blood Groups: Their Inheritance and Racial Significance," 256.

[228]  See E. von Dungern and L. Hirszfeld, "Über Nachweis und Vererbung biochemischer Strukturen," *Zeitschrift für Immunitätsforschung* 4 (1910), 531–546.

especially because both were inherited. In 1924, explaining that he had been occupied with the matter for a "long period of time," Hirszfeld reiterated his original claim that physiognomy and blood type were not related.[229] German physician Wilhelm Sucker agreed. His research indicated that the people of Leipzig were very similar in their blood type distribution to the Balkan peoples examined by the Hirszfelds. The two groups, however, generally "looked different," with the Germans tending to have much lighter pigmentation than individuals from the Balkans.[230] American researcher Grove similarly observed that one racial type of subjects, the Ainu, had the same blood type index as the "Senegalese Negroes," a group regarded by anthropologists as an altogether separate "racial type." The reverse set of these circumstances could occur as well; groups could appear similar but have very different blood type distributions. Portuguese physician A. Mendes-Corrèa referred to the Japanese and Chinese who, in spite of their different blood type indices, were known to fully resemble one another (*des affinités anthropologiques bien connues*).[231] To complicate matters even further, some studies revealed that even groups of the same nationality and "relative physiognomic homogeneity" could have widely divergent blood type distributions. Steffan remarked that the distributions of some groups in Central Europe, "in spite of their appearance," suggested that they were related to Northern Europeans. How the Alpine groups of Central Europe came to acquire "Mongoloid" physical characteristics, he did not know; the blood type index of "racially pure Mongols" did not indicate that they were closely related to the Alpine peoples.[232] Others simply pointed out that the racial types of A and B did not correspond to those categories already established by physiognomy.[233]

Even with such conflicting evidence, many felt that there were still reasons not to dismiss completely studies of blood and race. Their ideologies threatened, *völkisch* physicians were more apt to draw attention to the pos-

---

[229] Hirszfeld, "Die Konstitutionslehre im Lichte serologischer Forschung," 1180.
[230] Sucker, "Die Isohämagglutinine des menschlichen Blutes und ihre rassenbiologische Bedeutung," 487.
[231] A. Mendes-Corrèa, "Sur les prétendues 'races' sérologiques," *L'Anthropologie* 36 (1926): 440.
[232] Steffan, "Weitere Ergebnisse der Rassenforschung mittels serologischer Methoden," 387. Steffan explains that how the Alpine came to acquire Mongoloid bodily characteristics was unknown—it was "just as difficult" to explain why the North American Indians had these as well. The racial index of "pure Mongols" did not suggest a relation to the Alpine, however. For the time being, in contrast to Günther, Steffan did not label the Alpine as "Eastern." Ibid., 378.
[233] Mendes-Corrèa, "Sur les prétendues 'races' sérologiques," 439.

itive attributes of blood science. Blood type was recognized as a Mendelian trait, after all, and hereditary features were still the preferred means of racial classification. Extensive postwar studies further confirmed the inheritance of the blood types—by 1924 more than 600 families altogether had been studied.[234] Very few questioned whether blood type changed with environment.[235] Instead, most physicians agreed that blood type was a lifelong trait.[236] In fact, by 1924, the German courts had even begun to allow blood typing in resolving paternity cases and as forensic evidence. Legal use and recognition of the blood types was so reliable that some cases resulted in charges of perjury if blood type evidence indicated that a person had lied about their past sexual partners.

Accordingly, a contingent of blood scientists continued to maintain that the blood types could be used to study "racial origins and relationships."[237] Paul Steffan felt that the bulk of research revealed this to be the case, and even Ludwik Hirszfeld continued to maintain that blood science had proved useful in anthropological studies.[238] Furthermore, proponents of seroanthropology rightfully pointed out that it was a very new area of study—less than a decade old—and many theorized that the anthropological significance of blood would become apparent only after more detailed and thorough research and more cross-referencing between blood and other inherited traits. When the American blood scientist, Ella Grove, questioned the racial significance of blood because two disparate racial types shared the

---

[234] Snyder, "Human Blood Groups: Their Inheritance and Racial Significance," 237. This refers to international studies on inheritance. Felix Bernstein, a leading biomathematician, confirmed and amended the work of Hirszfeld and Dungern on ABO blood groups, describing a more exact manner of inheritance. The three-allele concept (A, B, and O) was found to fit the population and family data best. Gottlieb, "Karl Landsteiner, the Melancholy Genius: His Time and his Colleagues, 1868–1943," 21.

[235] Studies such as Harper and Byron, "Influence of Diet on Blood Grouping," *Journal of the American Medical Association* 79 (1922): 2222–2223. In 1925 German physician Groll mentioned that "some believed" that the protein in blood could change, after which point its racial value would be partially, or completely, lost. See Groll, "Blutuntersuchung zum Rassennachweis," *Berliner Tierärztliche Wochenschrift* (1925): 114. By 1923, many researchers had clearly shown that the blood groups were in no way affected by various factors, such as quinine, calcium, digitalis, salicylates, arseno-benzol, arsenic, mercury, ether, chloroform, morphine, protein therapy, serum and vaccine therapy, infectious and other diseases, pregnancy and childbirth, castration, sports, X-rays, radium, galvanization, and so on. Blood groups even remained constant after death. See L.W. Howard Bertie, trans., *Individuality of the Blood in Clinical and Forensic Medicine* (London: Oxford University Press, 1932), 45.

[236] Due to factors such as age, sex, vocation, disease, drugs, anesthesia, X-rays, climate or living conditions. See Snyder, "Human Blood Groups: Their Inheritance and Racial Significance," 251.

[237] Ibid., 233.

[238] Steffan, "Weitere Ergebnisse der Rassenforschung mittels serologischer Methoden," 390, and Hirszfeld, "Über die Konstitutionsserologie im Zusammenhang mit der Blutgruppenforschung," 428.

same biochemical index, her colleague, Clark Wissler, responded that the only way to prove the value of blood was "to take some definite anatomical character and correlate [the two]."[239] To Grove, the trait of blood type alone did not seem useful for racial classification, but she admitted that it might be useful when supplemented with other characteristics, such as "the color of the eyes, hair and skin."[240] Karl Landsteiner similarly agreed that blood type by itself could not be used to identify an individual's race; it had to be placed in the context of other anthropological traits. It was not possible, he observed, to make a system of races based simply on the size of the nose or any other one character, and the same was true of blood type; it was just "one factor" in classification.[241]

In addition, because of their newness, no specific protocol was established in the postwar flurry of blood type surveys. Seroanthropology attracted all types. Analyses were not conducted solely by serologists. One theorist commented that blood studies merited "closer study by competent anthropologists"—a remark clearly directed at medical amateurs who, curious about the new science, had set out to investigate it themselves.[242] As a result, the methodology employed in blood surveys was not consistent. Even the relatively simple procedure of typing blood could yield inaccurate results if performed incorrectly Also, no one knew how many subjects were needed to support a hypothesis, though most knew an inadequate sampling when they saw it. Even Ludwik Hirszfeld remarked on the irregularities in applications of research: the methods used were inconsistent, and the subject samples were sometimes too small to allow for larger generalizations.[243] In 1926 he called for continued (well-organized) seroanthropological research and even suggested extensive blood type surveys of primates across the globe to determine whether they had "different blood type distributions as humans do."[244] Hirszfeld firmly believed that blood science would contribute to some of the most pressing issues in anthropology.[245] He was not alone. Even with the volume of research and repeated reports

---

[239]   Grove and Coca, "On the Value of the Blood Group 'Feature' in the Study of Race Relationships," 91.
[240]   Ibid.
[241]   Ibid.
[242]   Snyder, "Human Blood Groups: Their Inheritance and Racial Significance," 257.
[243]   Hirszfeld, "Über die Konstitutionsserologie im Zusammenhang mit der Blutgruppenforschung," 434.
[244]   Ibid., 427.
[245]   For details on methodology, see ibid., 455.

of discrepancies, there were researchers who stubbornly clung to the belief that blood science had great untapped potential. With the correct guidance and supervision to ensure consistency between studies, they speculated that blood science might provide useful, hitherto-unknown details about race, mental illness, pathology, criminology, and other traits.

The German Institute for Blood Group Research, established in 1926, planned to address these demands. After this point, studies of blood and race would be more strongly marked by a *völkisch* presence and the contributions of Central and Eastern European scientists. Whereas the problematic nature of seroanthropology caused a decline in blood type surveys overall, particularly in Britain and the United States, the formation of the German Institute for Blood Group Research indicates the commitment and persistence of a select group of Central and Eastern European anthropologists to isolating the racial significance of blood.[246]

---

[246] "The founding program of the German Institute for Blood Group Research mentioned the following Central and Southeast Europeans as external members: Ioannis Koumaris, professor of anthropology at the University of Athens; Frigyes Verzár, director of the Physiological Institute in Debrecen; Lajos Méhely, professor of zoology at the University of Budapest; two Czechs, Oscar Bail and Jindrich Matiegka; and one Bulgarian, Vasil Mollov." Turda and Weindling, eds., *Blood and Homeland*, 9.

# CHAPTER III

## ORGANIZING SEROANTHROPOLOGY: THE ESTABLISHMENT OF THE GERMAN INSTITUTE FOR BLOOD GROUP RESEARCH

By the mid-1920s there were enough blood type surveys to suggest that some affiliation might exist between blood and other physical characteristics. *Völkisch* physicians tended to interpret type B blood as a marker of Eastern descent and, though there had been findings to indicate otherwise, they further linked this type to other inferior traits—such as inherited conditions, amoral behavior, or even propensity for disease. A group of researchers in these nations believed that seroanthropology might assist in meeting their racial and eugenic objectives.

Paul Steffan's postwar blood type studies of native Germans drew the interest of anthropologist Otto Reche and played a key role in the decision to create an organization focused solely on blood science. Reche and Steffan founded the German Institute for Blood Group Research, which, in spite of its name, was receptive to foreign membership and contributions. Steffan co-founded the institute, but Reche was its principal founder and chair. Reche's instruction and experience in the field of racial anthropology, as well as his wartime experiences, shaped his decision to form the institute and would continue to shape its priorities throughout its existence. Reche's *völkisch* tendencies, his racial stereotyping, and his career objectives were a product of his schooling, the intellectual environment of the time, and personal ambition. These helped to set the objectives for the German Institute for Blood Group Research and how its research agenda was proposed to state authorities.

## Otto Reche and Racial Anthropology

Otto Reche was born a Prussian citizen in 1879, in lower Silesia, on the border with Bohemia. His mother was born in Magdeburg, and his father in Upper Silesia.[247] He studied medicine at the universities of Jena and Breslau, where he was exposed to a range of subjects—including anthropology, ethnology, anatomy, zoology, geology, geography, comparative philology, and botany, but also, importantly, folk history and prehistory.[248] Reche was instructed by Ernst Haeckel, a recognized figure in German race science, who had once been an assistant to Rudolf Virchow, co-founder of the German Anthropology Association. Haeckel's *Natürliche Schöpfungsgeschichte* (Natural history of creation), published in 1868, represented the human species in a hierarchy from the lowest "racial types," such as the Papuan and Hottentot, to the highest, or the Caucasian peoples, which included the "Indo-German and Semitic" races.[249] His ranking of peoples was also evident during the polygenism/monogenism controversy; polygenists believed that the different human races had distinct origins, while monogenists claimed they came from common biological roots. Haeckel believed that orangutans were the primal ancestors of the Asians, and gorillas were the ancestors of the Africans, while Europeans had descended from chimpanzees—generally recognized as the most advanced of the primates.[250] Haeckel and Reche corresponded on many subjects, with an emphasis on zoology, biology, the origins of man and the different races—and the "Aryans" (*Indogermanen*) in particular.[251]

Reche had also studied with Felix von Luschan. Unlike Haeckel, Luschan was widely regarded as the leader of the liberal tradition in German anthropology after Virchow's death in 1902. In 1911 Luschan was appointed the first chair of anthropology at Humboldt University in Berlin. He also acted

---

[247] Geisenhainer, *Rasse ist Schicksal*, 23.

[248] Ibid.

[249] Robert J. Richards, *The Tragic Sense of Life: Ernst Haeckel and the Struggle over Evolutionary Thought* (Chicago: University of Chicago Press, 2007), 269. It is important to note that Haeckel did not support natural selection but instead believed in a Lamarckian inheritance of acquired characteristics. See Michael Ruse, *The Darwinian Revolution* (Chicago: University of Chicago Press: 1979). There is still discussion over whether Haeckel was a Darwinian or a Lamarckian; considerable debate also surrounds Haeckel's role in the rise of *völkisch* race science at the beginning of the twentieth century.

[250] Richards, *The Tragic Sense of Life*, 252.

[251] Geisenhainer, *Rasse ist Schicksal*, 46.

as director of the department for Africa and Oceania at the Museum for Ethnology (*Museum für Völkerkunde*) in Berlin.[252] Although Luschan did not make hierarchical claims comparable to those of his colleague Haeckel, he was similarly interested in racial categorization. Luschan became particularly well known for his creation of a chromatic skin pigmentation scale—one of the many tools designed to facilitate effective racial diagnoses. To determine the category of the subject, the examiner would compare that person's skin tone to one of Luschan's thirty-six opaque glass tiles.

How Haeckel and von Luschan helped to shape Reche's interests would be apparent throughout his career. In the summer of 1906, Reche was hired by the Hamburg Museum for Ethnology as a scientific assistant and director of its anthropological department. By the following winter, he was already giving lectures such as "Characteristics of the Human Races," "Geographic Expansion of the Human Races," and "The Influence of Environment and Culture on Physiognomy."[253] In the following year, Reche taught physiognomic measuring techniques in his first "Anthropometric Practicum." Reche provided instruction on this standard practice in racial anthropology, but he also aspired to improve existing means of racial analysis. In his early years at Hamburg, Reche closely examined skull and nasal indices of both humans and animals. Like the cranial index, the nasal index was simply another characteristic of the skull. Reliance on skull measurements stemmed from their supposed objectivity. In principle, measuring the skull and nose were simple. For each, low numbers corresponded with "inferior" racial types, while higher ones indicated "superior" racial makeup. Among his subjects, Reche found the lowest nasal index to belong to a fox terrier, and the highest to an individual from Hamburg—or "a representative of the Northern European race."[254] Reche deliberately compared animal and human indices; the placement of a dog at the opposite end of the "nasal spectrum" from the Northern European individual likened those racial types with lower nasal indices to animals—as with Steffan and his "Gondwanic" implication that non-European peoples were less evolved. Reche likely decided to focus on the nose instead of some other

---

[252] James Braund, "The Case of Heinrich Wilhelm Poll (1877–1939): A German-Jewish Geneticist, Eugenicist, Twin Researcher, and Victim of the Nazis," *Journal of the History of Biology* 41, no. 1 (2008): 10.

[253] Geisenhainer, *Rasse ist Schicksal*, 58.

[254] Ibid., 58.

trait because, given a choice, most racial analysts undoubtedly would have preferred measuring only the nose, which was considerably smaller than the skull and therefore less affected by the complexities of facial shapes.

Reche's first anthropological publication in Hamburg proposed a new and supposedly more efficient method of measuring the nose as a means of racial differentiation.[255] His proposal, however, failed to elicit any notable response from anthropologists. In the years leading up to World War I, Reche continued his employment in Hamburg, where he remained for the most part, aside from an anthropological fieldwork study in the South Pacific between 1908 and 1910 which, again, emphasized the importance of physical features in determining race.

In those years, Reche participated in the Hamburg South Sea Expedition, a study that planned a complete survey of the anthropological and ethnological traits of natives in Melanesia. The research took place in a militarized atmosphere, in which scientists often used weapons and threats of violence to force their subjects to hand over cultural artifacts or participate in measurements.[256] The Germans would often arrive in villages backed by armed escorts, as groups of antagonized islanders had attacked the anthropologists on several occasions.[257] They were able to collect the skulls and/or skeletons of about 800 subjects for the Hamburg museum. Josef Mengele took his first doctorate under Theodor Mollison, a member of the German Institute for Blood Group Research, with a paper on the anomalies of the jaw detected on Melanesian skulls which had possibly been collected by Otto Reche during his work in the South Pacific. Reche had been assigned to examine various physiognomic traits. He had skin and eye color charts, and he also recorded hair color. Of particular interest to Reche were the subject's head- and face-shape. He observed that the physiognomy of most of the subjects "was quite dull and unintelligent...their race impressed me as very ethnic and primitive" (*recht stumpf und wenig intelligent*).[258] Curiously, he further noted that some of the South Sea natives had an "almost Jewish appearance" (*ein fast jüdisches Aussehen*), from "their Jewish noses

---

[255] Geisenhainer, *Rasse ist Schicksal*, 58.

[256] Glen Penny and Matti Bunzl, eds., *Worldly Provincialism: German Anthropology in the Age of Empire* (Ann Arbor, MI: University of Michigan Press, 2003), 214.

[257] Ibid.

[258] Geisenhainer, *Rasse ist Schicksal*, 67.

to their Jewish profiles."[259] The results of this extensive research were published in the series *Ergebnisse der Südsee Expedition 1908–1910* (Results of the South Sea Expedition).[260] For his role, Reche was awarded a medal of honor from the city of Hamburg.[261]

During World War I, Reche took part in both active military service and in anthropological research. He was thirty-five when the war broke out in the summer of 1914; he was then commissioned as a lieutenant and sent to the front, where he was appointed major lieutenant early in 1915.[262] Reche suffered a gunshot wound and was consequently relieved of duty in October 1917, after which he joined a racial study on prisoners of war.[263] Like the circumstances under which the Hirszfelds conducted their research in Salonika, the wartime confinement of diverse peoples in other areas of Europe provided ideal conditions for anthropological research. Many anthropologists in these nations regarded the captives held in prisoner of war camps as prime "material" for scientific research.[264] However, in contrast to the Hirszfelds' study, these other analyses, many of which were in Germany and Austria, were focused mainly on physiognomy. These studies were often initiated and financially supported by the state; in 1915 a commission was established in Germany to gather information on different racial groups in prisoner-of-war camps. Reche was involved in one such study that arranged for the examination of several thousand captured soldiers in camps in both Austria-Hungary and Germany.[265] Thousands of data sheets were compiled on the subjects, in addition to the hundreds of photographs, plaster casts of heads, and hair samples collected.[266] Felix von Luschan helped coordinate the study, working with its director, anthropologist Rudolf Pöch, to standardize physiognomic measurements.

---

[259] Ibid.

[260] Georg Thilenius, ed., *Ergebnisse der Südsee-expedition 1908–1910* (Hamburg: Friederichsen, De Gruyter & Co.m.b.H., 1936). See also Otto Reche, "Untersuchungen über das Wachstum und die Geschlechtsreife bei Melanesischen Kindern," *Korrespondenzblatt Deutsch Gesellschaft für Anthropologie* 41, no. 7 (1913): 49.

[261] Berenbaum and Peck, eds., *The Holocaust and History*, 120.

[262] On August 1, 1914, he was named lieutenant of militia , then appointed lieutenant on February 9, 1915. See Geisenhainer, *Rasse ist Schicksal*, 94.

[263] Reche was shot in the chest, and this injury caused an "innervation" of the heart, which affected its beat (ibid., 94). See also Margit Berner, "From 'Prisoner of War Studies' to Proof of Paternity: Racial Anthropologists and the Measuring of 'Others' in Austria," in Turda and Weindling, eds., *Blood and Homeland*, 44.

[264] Ibid., 42.

[265] Ibid.

[266] Ibid., 43.

Reche's preoccupation with European racial types, in particular the racial identity of Germans, is apparent throughout his analyses. The subjects Reche examined were primarily of Central or Eastern European descent. Despite his subject groups' relatively close geographic proximities, Reche emphasized distinct racial differences among them and often drew attention to those individuals of supposedly superior racial stock, or those with German racial traits. Estonians, Latvians, and Lithuanians, he explained, possessed an extraordinarily strong proportion of Northern European blood, while the Dutch were "anthropologically German."[267]

After the war, Reche remained in Hamburg until 1924, when he was named professor of anthropology and ethnography at the University of Vienna. In December of that year he was appointed vice president of the Vienna Institute of Anthropology.[268] Here Reche repeated lectures on anthropometric practice consistent with his previous work. His subject matter and methodology did not noticeably shift until 1926, when he was almost certainly first introduced to the concept of seroanthropology. In that year Paul Steffan presented the results of his blood type surveys to the Vienna Institute of Anthropology. Steffan's research not only proposed a more efficient method of racial classification—which would have piqued Reche's interest, as demonstrated in his previous work with nasal indices—but also focused mainly on the racial makeup of Germans. Steffan's interpretation of differences in blood type distribution, and even type itself, demonstrated the same bias as Reche. Steffan attributed biological reactions to racial antagonisms. Although it was long understood to be the natural consequence of incompatibility between types, the "clumping" or agglutination of blood was attributed by Steffan to a more sinister cause. He claimed in 1923 that

> the agglutination of red blood cells to a foreign blood type (serum) suggests that these cells are all equipped with immunity against this type. Immunity is always formed in response to a threat—similar to reactions against parasites or poisons. Therefore, when one blood type agglutinates certain cells, but has no influence upon others, this confirms to us that the cells of some men in relation to others are perceived as foreign...[269]

---

[267] Ibid., 44.
[268] Geisenhainer, *Rasse ist Schicksal*, 113.
[269] Spörri, "'Jüdisches Blut,'" 36.

Steffan intentionally likened incompatible blood types to "parasites, poisons, or a foreign, bacterial source."[270]

Reche, who had expressed a mutual interest with Steffan in identifying and preserving the "German race," proved to be a receptive audience. This seemed to be the case especially after the war, when Reche's political activities revealed his even more pronounced concerns for the German *Volk*. The already low German birthrate had declined even further, and those Germans who were racially "most valuable" were also having the fewest children. Even the German peasants, Reche exclaimed, the healthiest racial stock, were practicing birth control.[271] Reche had also expressed concern about the effects of miscegenation upon the German people. Steffan's research, which sought to determine the advance of Eastern type B blood into Germany, was directly relevant.

With his own training and experience in predominantly physiognomic methods, Reche relied upon them as well, but he was also familiar with their drawbacks. The physical indicators of race were often easy to recognize, Reche explained, and one could even use a photograph or portrait if a physical exam was not possible, as these revealed the "color of the skin, eyes, and hair, as well as the shape of the head, face, and hair."[272] In theory, this was simple. For instance, Reche would categorize a family with light pigmentation as Northern European and "racially Germanic."[273] Nonetheless, he cautioned against relying solely upon such variables. While Germanic in appearance, this same family also had "Mediterranean blood," a trait which would not have been evident from a mere image, but was clear in their "lively, passionate temperament and heightened sensuality."[274] Racial mixing such as had occurred in this family compromised accurate racial categorization. In such instances, Reche believed it necessary to examine the entire family—both children and adults. To avoid mistakes, Reche also advised repeatedly measuring children throughout the course

---

[270]  Ibid. Similar analogies—again, though incorrect—were often repeated in the blood group research, despite Italian serologist Leone Lattes's report in 1923 that "at the present time, the view that iso-agglutination is related to disease of any kind is definitely abandoned and is only of historical interest." Leone Lattes, *L'individualità del sangue*, translated into English as *Individuality of the Blood in Biology and in Clinical and Forensic Medicine* (Oxford University Press, 1932).

[271]  Geisenhainer, *Rasse ist Schicksal*, 121.

[272]  Ibid.

[273]  Ibid.

[274]  Ibid.

of their development, as external characteristics could alter with time. Even with such precautions, however, he acknowledged that racial classification of a "mixed people" could be a useless pursuit, as "almost every phenotype could be a falsification of the actual facts."[275]

Reche had also remarked on the intricacies of physiognomic research during his prewar fieldwork in the South Pacific. Measuring subjects was often time-consuming, as anthropologists not infrequently undertook detailed physical examinations with numerous calipers and color charts. To expedite the process, Reche used specific hair and eye color charts in his evaluations. Nonetheless, he still had difficulties examining so many individuals quickly and efficiently.[276] In addition, Reche's analyses were further delayed by the need for interpreters, who were often employed to explain to foreign subjects the purpose of the exam and the measuring tools.[277] Reche complained about having to share one interpreter:

> It grows increasingly obvious that we only have one interpreter. Naturally, Dr. M. has him almost the entire day for his language studies and, without the interpreter, it is often almost impossible to measure and photograph the people. It takes at least three times as long otherwise, and I have so many who are afraid of the [measuring] instruments that they run from them. Of course, Professors F. and Hellwig need the interpreter as well. It has turned into a daily struggle.[278]

Testing an individual's blood, by comparison, was much simpler. Unlike the range of anthropomorphic racial criteria, blood type was objective. It did not change over time, or with a different environment, and its inheritance was generally understood. There would be no need to re-examine subjects periodically, as Reche suggested with children, as blood type did not change; it was set by the first year of life. Rare medical reports of blood types that had changed within an individual were disregarded by the German medical establishment and simply attributed to flawed research. By contrast, physical traits were affected by environment, even in adults.

---

[275] Ibid., 115.
[276] Ibid., 67.
[277] Ibid.
[278] Ibid. (Reche commenting on his anthropological expedition in the South Seas, Melanesia, August 1908.) Reche is likely referring here to Wilhelm Müller, who accompanied the expedition as an ethnologist and linguist; the other researchers are Friedrich Fülleborn and Franz Emil Hellwig. See Ibid., 63.

Reche was familiar with research, notably that of German anthropologists E. Hahns and Eugen Fischer, which suggested this to be the case. Their work prompted Dr. Robert Stigler, a participant at the 1926 meeting of the German and Viennese Anthropological Society, to call for an explanation of the inconsistencies in racial physiognomy. [279] Steffan agreed, claiming that anthropology had been "misled for decades" by the cranial index, which often resulted in "arbitrary and false" results. [280] Furthermore, blood science had already proven useful in studies of heredity, human biology, eugenics, and criminal anthropology. All of these factors served as an impetus for Reche to pursue seroanthropology. Steffan had hoped to elicit a response with his research, and Reche was willing to support him.

## THE GERMAN INSTITUTE FOR BLOOD GROUP RESEARCH

Shortly after their first meeting, in the summer of 1926, Steffan and Reche established the German Institute for Blood Group Research. Reche acted as chair. (See Figure 5.) They submitted an application for recognition to the Department of the Interior of the Chancellery in Vienna. The statement of purpose was rather general: the institute planned to conduct "uniform and precise" anthropological blood type surveys throughout Austria, Germany, the rest of Europe and, if possible, other continents. Relevant scientific contributions from outside the institute would be considered as well. Reche voiced his hope that the institute would resolve two important issues: whether there was a connection between serological type and other racial characteristics, and the precise nature of the inheritance of the blood groups. Responding to either would first require more extensive (and costly) research, but Reche reminded them that other disciplines, namely "immunology, eugenics, genealogy, and anthropology," stood to benefit from such work. [281]

In October 1926, seeking state assistance, Reche submitted a letter to Württemburg's Medical Research Office. The institute proposed typing the blood of 500 children in twenty-two districts. Reche knew that this was no small request and prefaced it by emphasizing the value blood might

---

[279] Otto Reche, "Zum Geleit," *Zeitschrift für Rassenphysiologie* 1 (1928/1929): 2.

[280] Steffan, *Handbuch*, 394.

[281] Ibid., 130.

have in racial classification. The first priority of the institute, he explained, was to determine whether the blood types were "in some way related to the human races—as the Hirszfeld brothers noted." The erroneous reference to the Hirszfelds indicated how little he knew of the field; this was further borne out by a call for more "extensive statistical surveys" in order to determine the nature of the relationship between blood and race.[282] In fact, the list of international seroanthropological studies at this point was fairly extensive.[283] Reche loosely referred to preliminary seroanthropological research—first with the Hirszfelds in Salonika, followed by that in the United States, Sweden, Finland, Poland, and Hungary (curiously, he did not mention the large contingent of Asian blood scientists). Many of these nations had gone beyond the beginning stages of seroanthropological research, to the point that, disillusioned by inconsistent results, they had already decided to abandon it altogether.

Reche's claim that blood type surveys had only "just begun" in Germany and Austria was accurate. In spite of the postwar boom in blood type surveys, statistics on Germans were relatively lacking due to poor funding and coordination. Accordingly, the first priority of the German Institute for Blood Group Research was a large-scale survey of the "native peoples" of Germany and Austria. Reche explained to the Württemburg officials that this would be possible only through government support. Continuing his plea, he pointed out how state funding had enabled impressive anthropological accomplishments in the past, specifically the work of anthropologist Rudolf Virchow. Hoping to determine the racial lineage of the German people, in the nineteenth century Virchow had recorded the physiognomic measurements of 6.76 million children—the largest racial study ever conducted.[284] Reche emphasized how "extremely important" the study had

---

[282] E151/54, 377. Reche eventually became aware of the fact that the Hirszfelds were a married couple, as he later referred to them as such; see Otto Reche, "Blutgruppenforschung und Anthropologie," Volk und Rasse 3, no. 1 (1928): 6.

[283] For a complete list of literature on blood science between 1901 and 1931, see Michael Hesch, "Das gesamte Schrifttum der Blutgruppenforschung in den drei ersten Jahrzehnten ihrer Entwicklung, 1901–1933," in Steffan, ed., Handbuch, 539–646.

[284] Gregory Paul Wegner, Anti-Semitism and Schooling under the Third Reich (London: Routledge Falmer, 2002), 12. Over the course of several years, the children's eye, hair, and skin pigmentation, and skull indices were recorded. See Rudolf Virchow, "Beiträge zur physischen Anthropologie der Deutschen, mit besonderer Berücksichtigung der Friesen," Abhandlungen der königlichen preussischen Akademie der Wissenschaften, Physisch-mathematische Klasse, Abt. I (1877): 1–390. Virchow's article was translated into English and published as "Blondes and Brunettes in Germany," Science 7, no. 157 (1886): 129–130.

Figure 5. Otto Reche. From Katja Geisenhainer, "Rasse ist Schicksal":
Otto Reche (1879–1966); Ein Leben als Anthropologe und Völkerkundler
(Leipzig: Evangelisches Verlaganstalt, 2002).

been and how it was made possible by the financial backing of the German Ministry of Education. What Reche did not mention in his plea was that Virchow concluded that racial uniformity did not exist anywhere in the Reich.[285] The results were so inconsistent as to lead Virchow to discredit racial theory. Furthermore, many Germans had criticized the study as a wasted effort and poor use of government funds.

Despite critical overtones, interdepartmental correspondence indicated interest in Reche's proposal.[286] The officials in Württemburg were apparently not concerned about the anthropological objective of Reche's research; studies in racial anthropology were commonly funded by the state. The authorities did, however, question the feasibility of the research; Reche's calculation that fifty blood samples could be collected in an hour seemed to them a "gross overestimation." They believed that even under ideal conditions, with the necessary disinfection, perhaps only ten could be finished in the space of an hour—which meant that it would take at

---

[285] Wegner, *Anti-Semitism and Schooling under the Third Reich*, 12.

[286] E151/54, 377.

least "six days of continuous work" to perform 500 tests. This difference in estimates was significant. The officials found it unlikely that the serological tests could be finished within the proposed timeframe. If it proved to be more time-consuming—as they felt was likely—it would require more funding. Nonetheless, the officials were sufficiently interested to recommend that Reche be accommodated as, they reasoned, blood science was still a relatively new area of research that might prove "quite important." The Württemberg medical authorities recommended that districts with full-time medical officers assist in the project.[287]

Reche submitted similar letters of request to state authorities throughout Germany and Austria. The Reich Ministry of the Interior reviewed the proposal and, like officials in Württemburg, also expressed reservations about the pace of the research.[288] They too found it unlikely that fifty blood samples could be taken in an hour, but this concern paled in comparison to their larger misgivings about the project. Because it was recognized by their courts for forensic purposes as early as 1926 the authorities in Prussia would have been much more familiar with the clinical procedures and technicalities of blood science than those in Württemburg. For them, Reche's letter raised many questions. Even if the blood was efficiently collected, how could they be certain that it would be typed properly? To ensure complete accuracy, should not each individual have his or her blood type tested more than once? This was standard procedure when using blood as forensic evidence. The matter of test serum was apparently a source of unease as well. Test serum, a byproduct of human blood, was a necessary component in blood typing. Because the serum was an organic material, the Prussian officials were concerned that it was "subject to change," which could lead to errors in blood typing. The only way to prevent such errors, they explained, was to employ "expert serologists," who were trained in the most accurate methods.[289] Having voiced these reservations, the Ministry

---

[287] Stuttgart, Ulm, Heilbronn, and eight rural districts did assist in acquiring blood samples, and the results were processed at Tübingen. The Ministries of Justice and the Interior supported the plan because of its potential value in serological testing for paternity and criminal cases. See Weindling, *Health, Race and German Politics*, 466.

[288] E151/54, 377. Reich Minister of the Interior, January 22, 1927, to the Provincial Government (*Landesregierung*) of Prussia.

[289] For further discussions concerning methodological errors, the author refers to the work of Schumacher and Artzerodt, "Fehler und Gefahren bei der Bestimmung der Blutgruppen," *Klinische Wochenschrift* (1926): 2016. The Italian serologist Lattes believed errors in blood group determinations to be an important cause

of the Interior believed it unlikely that the Reich Health Office would assist Reche. Reche's proposal, which had been processed with such relative ease in Württemberg, met instead with apprehension in Prussia.

The difference in responses can be attributed to the two institutions' respective experience with racial science. Simply put, the Reich Health Office was more familiar than the Württemburg Medical Research Office with blood research. The Reich Health Office's correspondence indicates that its officials were well-versed in the subject. Their misgivings and doubts about Reche's research came from a closer knowledge of the complexity of serology. By 1927, when the Reich Health Office received Reche's letter, blood typing was routinely being used in regional German courts. In paternity disputes, blood type could be used to rule out a suspected father. In criminal cases, typing the evidence, such as blood spatter on a shoe, could help determine a suspect's involvement. The blood in question was typed more than once, not only because of the seriousness of the charges—which could result in alimony payments, lengthy imprisonment, or even capital punishment, but also because there had been occasional reports of inconsistent results. For instance, a putative father might be identified as type A in the first testing and type B in the second. The need for reliable results created a new medical/legal niche filled by the so-called "expert serologist"—an individual schooled in the most recent and foolproof methods of blood science. This is why the Ministry called for employing expert serologists and typing a sample repeatedly. The Reich Health Office believed that these measures were necessary to ensure accuracy. In responding to Reche's proposal, officials in Württemburg made no comparable references to any potential inaccuracies or inconsistencies posed by serological research. No mention was made of expert serologists. On the contrary, the Württemburg authorities had suggested commissioning general physicians in the surveys.

To decide whether to fund the German Institute for Blood Group Research, the Prussian Health Council formed a committee to discuss the possible racial significance of blood. Coincidentally, Heinrich Poll, who had also been the director of the World War I prisoner of war study in which

---

of discrepancies: "The possibility is clear. Even in the particularly accurate determinations carried out in the Mayo Bros. Clinic, errors have occurred: out of 1,000 determinations studied by Pemberton, nine were found to be erroneous." Lattes, "L'individualità del sangue," 51.

Reche took part, was responsible for evaluating the proposal. Poll advised caution in funding what was still a "controversial area of research" because of its newness.[290] Hesitation elsewhere stalled Reche's plans in Württemburg. Officials there had yet to give a definite yes and explained their delay to Reche in October 1926:

> This decision [to fund the research proposal of the German Institute for Blood Group Research] is not being taken lightly. Accordingly, the Reich Health Office has formed a committee to further discuss research of the blood groups and advise accordingly. The Ministry feels it necessary to await the advice of this committee prior to fulfilling the request of the German Institute for Blood Group Research.[291]

The Prussian officials, however, had already denied the grant in March of that year. Shortly thereafter, Reche's plans were abandoned by the Württemberg authorities, and further rejections followed from Saxony, Bavaria, and Baden.[292]

These decisions revealed how little interest and confidence the larger German medical community had in racial applications of blood science. This was further verified by a decision made by the Emergency Association in late 1927. Friedrich Schmidt-Ott, president of the organization, invited a number of well-known anthropologists and geneticists to a meeting, along with some officials of the Reich Ministry of the Interior.[293] "For some time now," his invitation began, "the applications for support in the area of blood group research, of race research, and of anthropological studies have been increasing," adding that "with this abundance of individual applications," it was often difficult to select which projects to fund.[294] The purpose of the meeting with the invited specialists was thus to outline "the existing research tasks in this area" and sound out "the possibilities of working on them with the most thorough possible exploitation of the available funds."[295] The meeting took place on December 17, 1927. Among the par-

---

[290] Weindling, *Health, Race and German Politics*, 466.
[291] E151/54, 377.
[292] Weindling, *Health, Race, and German Politics*, 466.
[293] Hans-Walter Schmuhl, *The Kaiser Wilhelm Institute for Anthropology, Human Heredity, and Eugenics, 1927–1945: Crossing Boundaries* (Dordrecht: Springer, 2008), 84.
[294] Ibid.
[295] Ibid.

ticipants were the leading genetic researchers Erwin Baur, Carl Erich Correns, Richard Goldschmidt, Hans Nachtsheim, and the anthropologists Eugen Fischer, Theodor Mollison, and Otto Reche, along with the university lecturers Walter Scheidt and Karl Saller.[296]

For his part, Reche had hoped to make blood group research the main focus of the research agenda. While a number of the participants were interested and had even conducted blood type research themselves, they did not share Reche's enthusiasm for the science. Fischer remarked that, as seroanthropology recorded just one single hereditary attribute, one might "just as well support nose research."[297] In the year just before Fischer's proposal, the Emergency Fund had voiced its criticism of blood group research for "anthropological, racial and constitutional studies."[298] Fischer's "instinctive feeling" that the blood types had nothing to do with race was instrumental in the association's decision. Oddly, Reche even lacked the support of some in his own camp. Eugen Fischer was a member of the German Institute for Blood Group Research and had previously acknowledged that "in Central Europe, serological differences appear to exist between the Nordic, Alpine, and Mediterranean racial types."[299] Paul Steffan would later complain to Reche about Fischer's lack of interest, which had been clear to him during a visit to Freiburg in 1922 or 1923. At that time, Fischer had calmly stated that he "didn't think much of the blood groups" (for anthropological study).[300] The conference in 1927 reiterated this sentiment, declaring itself against mass statistical studies of cross-sectional groups like army recruits or schoolchildren—the very groups that Reche's institute proposed examining.[301]

Before the German Institute for Blood Group Research was even two years old, Reche already found himself grappling with repeated rejections of his proposed plans and efforts to attain financial support.[302]

---

[296] Ibid.

[297] Ibid., 85.

[298] Weindling, *Health, Race and German Politics*, 466.

[299] As quoted in Siegmund Wellisch, "Blutsverwandtschaft der Völker und Rassen," *Zeitschrift für Rassenphysiologie* 1, 1928: 21–34. In a letter dated November 23, 1926, to his colleague Oswald Streng, Otto Reche complained that Fischer was "not really to be convinced…because he has the 'instinctive feeling' that blood groups do not have anything to do with race." Geisenhainer, *Rasse ist Schicksal*, 132.

[300] Ibid., 172.

[301] Schmuhl, *The Kaiser Wilhelm Institute*, 84. The conference was for compiling the selective, tendentially complete genealogic-genetic biological records of isolated populations.

[302] Geisenhainer, *Rasse ist Schicksal*, 132–133.

The authorities' responses cannot be attributed to financial difficulties only; the biological sciences prospered during the Weimar Republic, even during the national economic crisis.[303] By 1933 Germany, along with Austria, had collected more Nobel prizes in medicine than any other country in the world—and more than France, Great Britain, and the United States combined.[304] Generous funding was provided for racial anthropology. In 1927 the Kaiser Wilhelm Institute was established; it became the premier race-science organization in Weimar and National Socialist Germany. Fittingly, given the pattern of indifference with which Reche met, the topic of race and blood types was dealt with in only one paper produced at the institute before 1933: in 1929, external staff member Max Berliner studied the blood of the cattle at the Institute for Animal Breeding in Berlin, and he concluded that blood type was not a reliable criterion for race determination.[305] As it had elsewhere, blood group research did achieve practical importance at the Kaiser Wilhelm Institute in connection with paternity tests, which were requisitioned by the Berlin courts starting as early as 1924.

The German state declined to fund seroanthropology to the extent that Reche had hoped. Instead, it financed the much more established study of physiognomic racial traits. The rejection of Reche's extensive survey plans of Germany and Austria created an opportunity for Eugen Fischer. In February 1928, at the Emergency Fund for German Science, Fischer presented a proposal for a national anthropological survey.[306] In spite of Virchow's study, which, it will be recalled, included over six million subjects, Fischer argued that little was known of the hereditary and racial composition of the German *Volk*. He planned to include studies of "stable, rural" groups to determine this.[307] The study was to include mainly physical traits, which were to be supplemented on occasion by blood type.

Among the characteristics considered "absolutely necessary" were the shape of the head, height, hair color, eye color, and the shape of the nose.[308] Examiners were instructed to take "as many photographs as possible, as

---

[303] Weindling, *Health, Race and German Politics*, 469. See also Aristotle A. Kallis, *The Fascism Reader* (New York: Routledge, 2003), 401.

[304] Since the introduction of the Nobel Prize in 1901. See Weindling, *Health, Race and German Politics*, 5.

[305] Schmuhl, *The Kaiser Wilhelm Institute*, 69.

[306] Weindling, *Health, Race and German Politics*, 466.

[307] Ibid., 466–467.

[308] Schmuhl, *The Kaiser Wilhelm Institute*, 85.

good as possible."[309] Blood type was included only in a secondary list of "highly desirable, but not necessary" characteristics.[310] Reche assisted with the latter.

Many parallels appeared to exist between Fischer's and Reche's proposals. Both were primarily concerned with identifying racially pure Germans. Both planned to examine rural, isolated peoples, and both proposed examining blood and physiognomy. However, while Reche's request for funding was denied, Fischer's was granted. The most glaring difference between their proposals was Reche's emphasis on blood instead of physiognomy. The state was more comfortable with physiognomy, the "normative methodology" of race science, and therefore more inclined to support it.[311] This preference for appearance in race had simple beginnings in the classical idea of beauty—the measure by which Europeans of the eighteenth and nineteenth centuries determined what was and was not visually pleasing[312]. Since this idea originated, race theorists created idealized images of beautiful races and juxtaposed these with stereotypes of racial ugliness.[313] Arbitrary preferences moved towards racial anthropology with the quanitification of beauty with Dutch anatomist Peter Campter's invention of the concept of the facial angle in the 1770s; the more the jaw protruded, the greater the angle, and thus, according to Campter, the more it resembled the angles of apes and dogs.[314] According to this system, "the Negro was more like the ape than like the Caucasian."[315] In Germany, the place where physical anthropology can be said to have been born, these ideas were very clearly articulated.[316]

As a racial anthropologist schooled in the German tradition, Reche was also taught to focus on these traits. This was consistent throughout his education and into his career as an instructor and researcher. This emphasis on physiognomy might have played into Reche's consistent reliance upon such indicators. Eventually, he would prepare careful guidelines for such

---

[309] Ibid.
[310] Ibid.
[311] Efron, *Defenders of the Race*, 11.
[312] Ibid., 14.
[313] Ibid.,
[314] Ibid.
[315] Ibid.
[316] Ibid.

criteria to be collected with blood type. At the same time, Reche apparently became frustrated with the nature of anthropomorphic research, which could be exhaustive in detail, as well as time-consuming, especially among those subjects believed to be "mixed." Even worse, he suspected that they were not fixed and could change, given the right circumstances. As a result, Reche's appeals for funding were denied, and the German Institute for Blood Group Research was forced to modify its original plans.

# CHAPTER IV

# SEROANTHROPOLOGY AT ITS HEIGHT: DISTINGUISHING THOSE WITH "PURE BLOOD"

Despite funding setbacks, the German Institute for Blood Group Research was able to pursue its research, albeit on a much less extensive scale than originally proposed. Although its directors welcomed analyses of non-German subjects, they were particularly interested in determining the serological makeup of what they termed "native Germans." Finishing the "mapping out" of the German people would remain one of the institute's main objectives, and throughout the late 1920s the institute would collect and examine the blood types of carefully selected German populations. Before 1926, institute co-founder Paul Steffan had been the first German physician to coordinate serological research with the purpose of determining the advance of Eastern type B blood into Germany. However, only minimal studies were conducted, causing Steffan to complain in 1925 about how incomplete research in this area was.[317] The institute would resume his work. In their search for the most "racially pure" Germans, studies would become increasingly intricate and often interdisciplinary, incorporating historical, archeological, and geographic details in their research. An overview of the research conducted by the German Institute for Blood Group Research demonstrates how certain authors were largely preoccupied with the racial implications of the different blood types. This again manifested itself in attempts to demonstrate an association between type B blood and racial or eugenic inferiority.

---

[317] Steffan, "Weitere Ergebnisse der Rassenforschung," 376–377.

The German Institute for Blood Group Research reported its findings in its periodical, the *Zeitschrift für Rassenphysiologie* (Journal for Racial Physiology), which began publication in 1928. Members of the institute shared in the larger discourse of European seroanthropology. In spite of the publication's title, which indicates that its contents were not limited to the serological branch of race science, the editors saw themselves as the center of a "national and international effort to study blood group distribution."[318] Therefore, both non-German membership in the institute and articles by foreign authors were common. For a work to be eligible for inclusion, Reche and Steffan were mainly concerned that studies present "original research." A close analysis of these studies—how they were planned, how their subjects were selected, and how results were interpreted—lends insight into the physicians' motivations.

## Studies of "Native Germans"

The German Institute for Blood Group Research was able to conduct studies of German schoolchildren through support from the Vienna Academy of Sciences, though the subject group was much smaller than the figure of 400,000 originally proposed. (See Figure 6.) Sharing Steffan's intent to distinguish the native Germans within the larger population, a group of physicians was similarly careful in selecting subjects. Schlossberger and his colleagues decided upon a number of villages near Frankfurt am Main, which they reasoned must occupied by descendants of the original occupants, a claim "supported by previous anthropological research and confirmed by a professor consulted at the Historical Museum of Frankfurt, as well as a teacher in Rendel."[319] To substantiate their decision, they referred to the area's long history; the names Rendel and Randwilre (Randweiler) were first mentioned in documents dating back to the year 780. The nearby communities of Grosskarben, Kleinkarben, and Carbah had been referred to

---

[318] William H. Schneider, "The History of Research on Blood Group Genetics: Initial Discovery and Diffusion," *History and Philosophy of the Life Sciences* 18, no. 3 (1996): 295–296.

[319] Hans Schlossberger et al., "Blutgruppenuntersuchungen an Schulkindern in der Umbegung von Frankfurt a.M.," *Medizinische Klinik* 24, no. 22 (1928): 851. The seven cities researched were Grosskarben, Kleinkarben, Rendel, Niedererschbach, Obererlenbach, Niedererlenbach, and Petterweil. The study includes a table that lists the specific blood type distributions of each location. See ibid., 852.

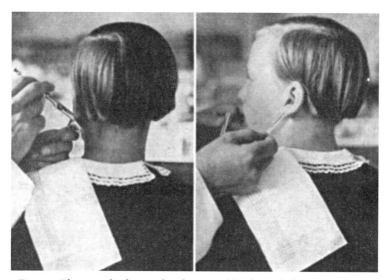

Figure 6. Photograph of researcher drawing a blood sample from a child's ear. Reche assured authorities that "thousands of schoolchildren" could be examined without danger to either the children or their hand; only two small drops of blood would be taken from their fingertip or earlobe. E.D. Schött, "Die Technik der Blutgruppenbestimmung" in Paul Steffan, ed., *Handbuch der Blutgruppenkunde*, (Munich: J.F. Lehmann, 1932), 470.

shortly thereafter, in 817.[320] The archeological evidence indicated that persons referred to as "native Germans" (*vorgermanischen Bevölkerung*) had been in the area even earlier, sometime after the middle of the first millennium BC. These peoples, particularly the Celts, had dispersed throughout western and southwestern Germany. These earlier settlers were ostensibly the ancestors of modern-day, "racially pure" Germans. However, the shifting power vacuums and population movements in the ancient world suggested that miscegenation had occurred between the native Germans and other peoples. The border of the vast Roman Empire had been close to this region, and the examiners reasoned that its various subjects had moved into Germany from regions as disparate as "Gaul, North Africa, and Asia." There seemed to be evidence of Roman influence. Primary sources from the eighth century AD mentioned Roman occupation. Historians claimed that the suffix "weil," as in the town of Petterweil, stemmed from the Latin word "villa." All signs suggested that the Rhineland had once been a Roman

---

[320] Ibid., 851.

settlement.[321] According to the evidence, the researchers expressed "no doubt" that their subjects descended from "a mix of different tribes and races." Further racial mixing had likely occurred even more recently—the result of an influx of refugees and immigrants brought to the area by wars, the Counter-reformation, and even the French Revolution, a conflict that had introduced "Walloons, Flemish, and French" groups. In addition, industrialization in the late nineteenth century attracted foreign workers, further complicating the mix.[322]

Amazingly, despite their own recognition of the areas' complicated ethnic history, and its modern-day results in a racially diverse people, the physicians believed that they had "singled out" those individuals least likely to have been affected by the circumstances of the past two millennia.

Miscegenation among their group, they declared, had been minimal; the "greater portion" was related to the region's original, pre-German inhabitants—and they could prove it. Church registries indicated that most of the subjects' families had been in the area as far back as the Thirty Years' War (1618–1648).[323] Further inquiries revealed that this was also the case with the children's parents; most of their fathers had been born in the immediate area, and while their mothers had often come from neighboring regions, such as Rhön, Vogelsberg, Odenwald, or Rheinhessen, it had only "rarely been the case" that they had lived far from German borders.[324] Of course, the glaring omission in their claims that these were "unmixed" subjects is that the examiners cannot account for possible (indeed, very likely) instances of miscegenation in the many centuries (largely devoid of sources relevant to tracing heritage) before the relatively recent Thirty Years' War. This was not addressed, though it was weakly mentioned that children of families drawn to the area by new industry had been "strictly avoided."[325]

---

[321] The authors explained that "many of these areas had originally belonged to the French—even Niedererlenbach originated in the time of Charlemagne." Hans Schlossberger, "Blutgruppenuntersuchungen an Schulkindern im Niedgau und in der südlichen Wetterau," *Zeitschrift für Rassenphysiologie* (1928/1929): 118.

[322] Ibid. In this area, about "half to two-thirds" of the inhabitants were textile workers. Ibid., 118.

[323] *Völkisch* race theorists often lamented the increased "racial mixing" that occurred after this point. These categories were almost always left undefined, serving more as shifting, prima facie cultural focal points rather than empirically demonstrated bases upon which experimental results could rest. This was a change also recognized by extreme-right political figures; in *Mein Kampf*, Hitler grieved the "poisonings of the blood which have befallen our people, especially since the Thirty Years War." Adolf Hitler, *Mein Kampf* (New York: Mariner Books, 1999), 396.

[324] Schlossberger, "Blutgruppenuntersuchungen an Schulkindern im Niedgau und in der südlichen Wetterau," 119.

[325] Ibid., 118.

Confident that their sample group served as an adequate representation of "native Germans," the examiners proceeded with their research. Their examination was thorough and included the usual listing of racial markers: the birthplace of each child's parent was recorded, as were his/her age, sex, religion, blood type, hair and eye color, and skull and face shape. With their light hair and eyes, most of the children conformed to the Nordic stereotype. Their blood types confirmed this: the children had an appropriately high level of "Western European" type A blood. Those with the "Eastern" type B were the exception.[326] For all practical purposes, it seemed to Schlossberger et al. that, although the odds had been against locating "unmixed" peoples, their selection process had been successful. The "Germanness" of the subjects had remained largely intact. Eastern traits, such as dark pigmentation and high levels of type B blood, were conspicuously absent.

Wilhelm Klein, a district medical officer in the Rhineland, performed a separate analysis of schoolchildren in the cities of Oberlahnstein and St. Goarshausen, who were also selected because they were believed to have descended from the "original inhabitants." While some of Klein's findings were consistent with this previous study, they also revealed indiscrepancies.[327] As was the case with the subjects in Frankfurt am Main, the children did have the expected "noticeably high percentage" of type A blood; it was so pronounced among individuals from Nassau (Rhineland) that Klein separately calculated the blood type distribution of those whose parents were native to that region; this group had even more type A blood and less B.[328] Klein was puzzled, however, by the lack of correlation between blood type and appearance. Even though there were more blond children over-all, which fit with the idea that this was a group of so-called native (Nordic) Germans, considerably more brunettes had type A blood—and high levels of type B, in turn, existed among the blonds. These results linked type A blood, that typically associated with Western European origins, with a dark, Eastern appearance. Klein came across further "inconsisten-

---

[326] 47.17 percent type A blood and 9.4 percent type B. Schlossberger et al., "Blutgruppenuntersuchungen an Schulkindern in der Umbegung von Frankfurt a.M.," 852.

[327] Wilhelm Klein, "Ergebnis der Blutgruppenbestimmung in Oberlahnstein und St. Goarshausen," *Zeitschrift für Rassenphysiologie* 1 (1928/1929): 12. Of the approximately 800 children in Oberlahnstein, 492 were examined, and ninety-two of the 150 children in St. Goarshausen were examined. Klein mentions that he was unable to obtain further children's blood group samples from elsewhere in the area, but he does not explain why.

[328] This group had 52.4 percent type A blood and only 8.3 percent type B. Ibid.

cies" when he compared skull shapes to blood types. There was a predominance of Nordic "long skulls" among both A and B types, but A was more common among individuals with the characteristically Eastern "round skull." Apparently under the impression that blood type and physiognomy were related, Klein referred to his findings as a phenomenon that had yet to be explained. Nevertheless, he pointed out that there were appropriately "more blonds with long skulls, and more brunets with round ones." Importantly, however, these were both visible differences, completely separate from blood. Klein concluded his analysis by referring to familiar physiognomic traits that confirmed that nothing was anthropologically amiss.

In reference to the discrepancies he came across, Klein entertained the notion that these were the result of what was, in actuality, a (racially) "impure" subject group—perhaps, the researchers mused, there had been miscegenation of which anthropologists were not yet aware. Unlike his colleagues in Frankfurt am Main, Klein believed that his subjects were quite likely "racially mixed," despite the general consensus that these settlements were "pure." Klein attributed this to the Rhineland's diverse history, previously described by Schlossberger et al. He acknowledged that the region had been traversed by various peoples (i.e., "racial types") serving as a "convenient passageway" over the course of the past millennium. Klein further blamed the Thirty Years' War and Spanish conquests for contact with non-German types; the Spanish War of Succession had definitely introduced a "certain measure of Spanish blood."[329] He referred to the village of Sauerthal, where many of the German inhabitants had either fled or were killed. It had then been settled by Spanish mercenaries, whose influence was still apparent in local Spanish surnames. Klein further theorized that subsequent shifting ownership of the lands—which belonged at one point to the clergy, then to various nobles, had also led to "strong mixing" (*starke Vermischung*). He came to a very different set of conclusions from Schlossberger et al. in their analyses near Frankfurt am Main. Instead of identifying native Germans, Klein argued that the area's tumultuous history had resulted in a "racially mixed" German peasantry—and this miscegenation was the source of the unexpected inconsistency between his subjects' blood type and appearance.

---

[329] Ibid., 14.

Paul Steffan and a colleague came to a similar conclusion based on their findings in Osnabrück, Germany. [330] They, too, examined blood types as well as various characteristics of 500 individuals. Osnabrück, too, was also chosen with the expectation that it would "still include many descendants of the native inhabitants" (*enthält noch viel alteingesessene Bevölkerung*). Steffan likewise defended this claim by an investigation of the area's history. In 783 AD, while under attack by the Saxons, Charlemagne moved the center of his diocese to Osnabrück. The city lay under foreign military occupation intermittently, beginning in the Middle Ages; Swedes were in the area for about a decade during the Seven Years' War—later came soldiers from both England and Hanover. More recent historical developments revealed patterns comparable to those in Frankfurt am Main. The French occupied the area for a period of time. Osnabrück's booming iron and textile industries also attracted a large influx of workers throughout the Industrial Revolution; many came from nearby agricultural areas, but some had traveled from farther away (for instance, Harz or Böhmen). Unlike his colleagues, Steffan referred to the additional possibility of miscegenation between Germans and Jews—an unlikely occurrence, he explained, as "no Jews lived in the city between 1424 and 1800." [331] Although its overseers had taken precautions to ensure that their subjects were "racially pure," this study also had a major flaw. Basically, Osnabrück's history was irrelevant, as the majority of the subjects examined were not even native to the area, but were women from elsewhere employed there as medical assistants. Confronted with this discrepancy, the examiners decided to divide the women based on whether they were from rural or urban regions of Germany.

Hoping to glean some useful information from this data, Steffan reverted to his previously discussed theory of racial mixing in cities. He anticipated that the blood type distributions would reveal that rural peoples had remained untouched by the "modern traffic" characteristic of cities, and these expectations were met. [332] Subjects from urban areas were completely different from their rural counterparts. Women from the cities

---

[330] Helfriede Meyer and Paul Steffan, "Die Beziehungen zwischen Blutgruppe, Pigment und Kopfform," *Zeitschrift für Rassenphysiologie* 2, no. 2 (1929): 46–60. Most of the subjects were medical assistants employed in Osnabrück.

[331] Ibid., 57.

[332] Ibid.

had higher levels of blood type B and were also "less Aryan" in appearance, with their higher incidence of dark hair, skin, and eyes. By contrast, the subjects from agricultural areas had proportionately more type A blood, as well as a more "Nordic" physiognomy. It was clear to Steffan that the miscegenation believed to have been more frequent in urban areas manifested itself not only in the subjects' appearances but also in their blood type distributions. For Steffan, blood typing confirmed his theory that the "purest, untouched German blood" was to be found in isolated areas.

Schridde, a physician at the University of Heidelberg, sought to differentiate his research on native Germans by examining a group even more isolated in the Oden Forest, where a blood type survey had yet to be carried out.[333] Schridde reported that villages in the forest were occupied by peoples closely related to the original, "unmixed" (*unvermischter*) inhabitants. The chosen villages—Gumpen, Reichelsheim, Winterkasten, Schlierbach, Mossau, and Hiltersklingen—were "almost exclusively" occupied by farmers. Through information acquired from the towns' mayors and clergymen, Schridde confirmed that immigration to the area had been rare. Again using church registries, he boasted that he could trace his subjects' lineages as far back as that now-familiar benchmark of racial purity, the Thirty Years' War. Schridde took the additional precaution of only examining individuals whose parents and grandparents had both been born in the immediate area. With the exception of Hiltersklingen, Schridde was able to analyze approximately 30–50 percent of the residents in each village.[334]

Even with such careful screening, the results were frustratingly inconclusive. Schridde expressed "surprise" at the inexplicably high frequency of blood type O. Type A was elevated in certain villages but quite low in others. Schlierbach had only a "modest representation" of type A blood and shared similar percentages of type B with Mossau and Gumpen, but

---

[333] P. Schridde, "Über die Blutgruppenzusammensetzung in einigen Odenwalddörfern mit altangesessener Bevölkerung," *Zeitschrift für Rassenphysiologie* 2 (1929/1930): 62–72. Schridde was employed at the Hygiene Institute of the University of Heidelberg, under the direction of Dr. Emil Gotschlich. The Oden Forest, or Odenwald, is a wooded region in Germany about 80 miles long and 25 miles wide, situated mainly in Hesse. It is also referred to as the "Badisch-Sibirien" (Siberia of Baden) because of its isolated location. Its remoteness and harsh climate retarded its cultural and economic growth for centuries. See Winifred Wackerfuss, ed., *Beiträge zur Erforschung des Odenwaldes und seiner Randlandschaften* (Breuberg-Neustadt: Breuberg-Bund, 1982).

[334] Schridde, "Über die Blutgruppenzusammensetzung in einigen Odenwalddörfern mit altangesessener Bevölkerung," 62.

these did not match the distributions in the other villages.[335] There was no discernible pattern to the blood type frequencies. Although isolated, the "untouched villages" did not resemble many other similarly remote locations in their blood type percentages—not even the "ancient farming areas of Oberhessen or Schleswig-Holstein."[336]

While there had been blood-type findings to suggest that the Germans were, in fact, "racially mixed," Steffan and other biased physicians continued to maintain that they had been able to identify descendants of the areas' supposed original inhabitants. This would have been reassuring to völkisch ideologues who feared that there was no longer a distinct race unifying the German people. Others had suggested this to be the case. Besides Rudolf Virchow's findings in the nineteenth century, German anthropologist Herman Wirth claimed that the German *Volk* had been originally composed of "Celtic, Lithuanian, Germanic and Slavic blood, then Romans, Jews, Huguenots and Italians, with a smattering of Swedes, Scots, Croats, Irish, Hungarians, Spaniards, and Turks."[337] In 1928 liberal politician and psychologist Willy Hellpach, the Democratic Party's presidential candidate in 1925, rejected the idea of a racial essence peculiar to the German character—there were too many ethnic types in Germany for the nation to be identified "with a single race or blood type."[338]

In their efforts to define the racial makeup of the German people, blood scientists did not always take the time to select their subjects carefully prior to analysis. On occasion they simply took advantage of easily accessible groups, such as military personnel and hospital patients. In 1929 M. Reich, an army physician, conducted anthropological examinations on over 600 soldiers, most of whom were from East Prussia.[339] As with other studies, this one was very thorough; the men's blood types were recorded, as well as their head and eye shapes, surname, and mother's maiden name.[340] On average, the incidence of type A blood in this group was much lower and type B was higher, compared to analyses of German subjects farther west.

---

[335] Ibid., 72.

[336] Ibid., 65.

[337] Hutton, *Race and the Third Reich*, 20.

[338] Weindling, *Health, Race and German Politics*, 482.

[339] M. Reich, "Soldatenuntersuchungen nach Blutgruppen und Konstitutionstypen," *Zeitschrift für Rassenphysiologie* 1 (1928–1929): 147–150.

[340] Ibid., 147.

Their distributions still classified them as "Western European" according to Hirszfeld's biochemical index, and Reich noted that type B was still relatively rare, but overall their serological profile more closely resembled that calculated for nearly 12,000 Poles than it did western Germans.[341]

When blood type statistics proved disappointing in their implication that the German people were "less Nordic" than was hoped, Reich, like his *völkisch* colleagues, turned to physiognomic traits to confirm the Aryan identity of the German people. He reported that type A blood was fittingly less common among the subjects with both "non-German" names and "Mongoloid-shaped" eyes.[342] It had also been more frequently reported among the subjects with "long skulls" and less so among those with "round" ones.[343] Because of their "inferior status," Reich also predicted that individuals with type B blood would be more susceptible to disease than those with type A. However, he reasoned that testing this would be difficult, as his subject group of soldiers consisted mostly of "strong, resistant" men.[344] This did not appear to have adversely affected his results, though, because just as he had anticipated, "many more" of the sick men had type B blood than A.[345]

The director of the Medical Research Office in Gumbinnen (Prussia), G. Schaede, expressed reservations about Reich's study, in particular the "unexpectedly high" levels of blood types AB and O reported. Based on these, he speculated that the research had been flawed. Critical of the homogeneity of Reich's subject group, Schaede decided to conduct his own study with a more appropriately diverse range of subjects.[346] At 3,000 subjects, Schaede's subject pool was considerably larger than Reich's. The individuals were Prussians from the towns of Gumbinnen and Allenstein, as well as "settlers from the Reich" and those "who had mixed with peoples from Salzburg." Their surnames indicated German but also Lithuanian and

---

[341] Thirty-eight percent of the soldiers were found to have type A blood, and 13 percent had type B. At 2.08, their biochemical racial index technically grouped them as "Western European" in type. Ibid., 149.

[342] These subjects had 29.4 percent type A blood, and 17.6 percent B. Ibid.

[343] Of those with Nordic long skulls, 41.7 percent had type A blood, which declined to 33.5 percent among individuals with round skulls. Ibid., 148.

[344] Ibid., 150. In spite of the fact that many of the men were endomorphic (muscular), Reich also examined body type in relation to blood group, but the number of subjects was too small to make any correlations.

[345] Of the men with type A blood, 54.6 percent were sick. This increased to 75.5 percent in the type B group. Ibid., 149.

[346] G. Schaede, "Blutgruppenuntersuchung in Ostpreussen," *Zeitschriften für Rassenphysiologie* 1 (1928/1929): 151.

Polish descent. Schaede recorded each subject's age, gender, and occupation. Type B blood was more elevated overall, though through more careful analysis, Schaede found that it was only higher among those with "Eastern characteristics." Type B blood was most common among individuals with Polish surnames who, Schaede theorized, "undoubtedly represented the average" (of the population in East Prussia).[347] However, the blood type distributions of those whose parents were born in East Prussia were found to be comparable to Germans farther west in Westphalia, who had the anticipated higher proportion of blood type A.

## BIASED RESEARCH

Studying the racial makeup of isolated German peoples addressed a principle central to radical German nationalism. Distinguishing native Germans, or those who represented the "original type" (*Urtypus*), from those who had mixed with "racial others" was a priority of far-right German racial anthropologists.[348] *Völkisch* ideologues had long theorized that the most "racially pure" Germans were the peasantry, those who worked the soil, isolated from the depravity and miscegenation presumed to characterize urban life. The greatness of the German *Volk*—biological and moral— came from the special connection between "Aryan blood" and Germany's soil.[349] "Blood and soil" became a rallying cry of the nationalist movement, which described aliens as not only lacking a connection with nature, but also as having no souls. By contrast, they emphasized the positive values of "true" Germans—their nobility, their connection to nature, the productivity of their work, and the superiority of their race. The best, "purest" group was the peasants—the workers and the creators—who were unique in caring for the soil and possessing "true culture" ("culture" and the German *Kultur* come from a Latin word for farming).[350]

Reche argued that higher capabilities and the gift of "cultural creativity" were most pronounced among the native Northern European racial types.[351]

---

[347] Ibid., 153. Schaede noted that these men were generally residents of East Prussia or had perhaps only been allowed to remain temporarily in East Prussia for "research purposes."

[348] Efron, *Defenders of the Race*, 102.

[349] George Victor, *Hitler: The Pathology of Evil*, 136.

[350] Ibid.

[351] Geisenhainer, *Rasse ist Schicksal*, 119.

The evil, "racially inferior" groups were the nomads and merchants.[352] The nomads exploited the soil without caring for it until it was used up, ruined.[353] Considering work a curse, they had no culture except what they took from indigenous peoples.[354] Merchants, the men of the city, were much like the nomads in lacking culture and valuing only money—they also exploited the industry and creativity of peasants.[355]

Other authors publishing in the *Zeitschrift für Rassenphysiologie* similarly believed that the most racially pure Germans existed in rural enclaves. Their studies were implicitly critical of the metropolitan growth that had occurred in Germany since the nineteenth century. To *völkisch* theorists the process of building up industry and cities seemed merely "materialistic," and it was destroying the old pastoral Germany, an "essential idyll in the romantic celebration of German identity."[356] From the fin-de-siècle period into the Third Reich, the image of the German peasant—strong and hard-working—was repeatedly contrasted with that of the Jew—cosmopolitan, uprooted, and foreign.[357] Hostility was also directed against Slavic groups who had migrated to German cities seeking employment and, as researchers mentioned in their analysis, to the more remote agricultural districts as seasonal workers. Martin Staemmler, a Chemnitz doctor and advocate of sterilization, demanded family welfare measures in "support of the German peasantry."[358] State authorities agreed; the recognized German scientist Erwin Baur "fell foul of the authorities" when he slandered certain rural settlements at a meeting of the Prussian Health Council.[359] During the Weimar Republic, increased fears regarding miscegenation and eugenic decline led researchers to try "sorting out" racially pure Germans even more aggressively than they had previously. They sought to determine not only how much racial mixing had occurred, but also how a "contemporary race" was related to its original type.[360] The main objective was to make

---

[352] Victor, *Hitler*, 137.
[353] Ibid.
[354] Ibid.
[355] Ibid.
[356] Nicholas Goodrick-Clarke, *The Occult Roots of Nazism: Secret Aryan Cults and Their Influence on Nazi Ideology* (New York: New York University Press, 1993), 4.
[357] Geisenhainer, *Rasse ist Schicksal*, 136.
[358] Weindling, *Health, Race and German Politics*, 479.
[359] Ibid., 471.
[360] Efron *Defenders of the Race*, 105.

certain that the original German racial type still existed—then the focus would shift to ensuring its survival and increase. Reche further proposed measures to protect these "superior" racial elements of Northern European peasant stock; he recommended sterilization laws to prevent "reproduction of the less valuable." In conjunction with legislation, he further supported immigration restriction, which would help to prevent unwanted racial mixing.[361]

These same concerns were voiced in extremist propaganda. In 1928 race fanatic Richard Walther Darré published *The Peasantry as the Living Source of the Nordic Race* (Das Bauerntum als Lebensquell der Nordischen Rasse), which was followed in 1930 by *A New Aristocracy from Blood and Soil* (Neuadel aus Blut und Boden). Darré's work was a blend of zealous Nordicism and medicine. His dissertation had been on the domestication of the pig, and he used this to demonstrate how the degenerative effects of domestication were comparable to civilized society.[362] This led to a doctrine of racial hygiene based on breeding a (racial) aristocracy from "uncorrupted peasant stock."[363] Darré argued for an "incipient new German nobility," which "must once again become a vital source of thoroughbred leadership talent. It must have at its command mechanisms that retain time-tested blood in the hereditary line, [and] reject deficient blood."[364] Such mechanisms, he thought, were available from early German history and the role it assigned the free peasant farmer.[365] In ancient times, Darré claimed, "the word 'peasant' was an honorific and expressed the concept of personal freedom." The transmission of blood was "symbolically linked to the—eternally lit—fire on the hearth," based on a "specifically defined landed property."[366] There was an integral link between Aryan racial identity and the land. To Darré what mattered was shoring up the peasant farmer, in essence the same "native German" sought in so many blood type surveys, as the "source of German racial strength."[367] Darré's publications contributed heavily to National Socialist teachings and shaped the Nazis' *Blut und*

---

[361] Geisenhainer, *Rasse ist Schicksal*, 121.

[362] Weindling, *Health, Race and German Politics*, 475.

[363] Ibid.

[364] Ben Kiernan, *Blood and Soil: A World History of Genocide and Extermination from Sparta to Darfur* (New Haven, CT: Yale University Press, 2007), 425.

[365] Ibid.

[366] Ibid.

[367] Richard J. Evans, *The Third Reich in Power* (New York: Penguin, 2006), 421.

*Boden* (Blood and soil) doctrine, which blended ideas of racial purity and pan-Germanism.

Both of Darré's texts were published by J.F. Lehmann Publishing House in Munich, which also printed the *Zeitschrift für Rassenphysiologie*. J.F. Lehmann was one of the largest publishers in the early twentieth century and the single largest German publisher of works in the field of racial hygiene.[368] Founder Julius Lehmann's active participation in Germany's right-wing community and a close acquaintance with Otto Reche indicate that the journal fit with his racist and nationalist tendencies. During World War I, Lehmann published military tracts and in 1917 founded the nationalist *Deutschlands Erneurung* (Germany's Renewal), the "first German journal to give the question of race and racial hygiene its deserved and proper place."[369] Following the war, Lehmann continued to be an active participant in the German nationalist and race movement. After Christmas 1918, he spent several months in prison for plotting a coup d'état unifying the League of Pan-Germanists, the anti-Semitic Thule Society, and the Oberland League against the Republic of Soviets of Munich. Upon his release, Lehmann helped to organize the ultra-nationalist *Deutschvölkische Schutz- und Trutz-Bund* (German people's protection and defense league), which mobilized the public with its anti-Semitism and xenophobic belligerence.[370] In 1919 Lehmann was sent back to prison after one of the members of the Thule Society assassinated Kurt Eisner, leader of the Bavarian communists. He then rejoined the Freikorps headed by General von Epp, who reclaimed the city of Munich from the communists. Lehmann even played a role in the Hitler putsch in Munich in 1923 and continued the distribution of National Socialist material while Hitler was imprisoned at Landberg.[371]

Lehmann and Reche were colleagues before Lehmann became affiliated with the institute. When Lehmann launched the popular right-wing race journal *Volk und Rasse* (Nation and race) in 1926, he appointed Reche as editor-in-chief. Lehmann regularly contributed to the publications of racially biased periodicals such as *Volk und Rasse* and the *Archiv für Rassen- und Gesellschaftsbiologie* (Journal for racial and social biology); Lehm-

---

[368] Proctor, *Racial Hygiene*, 122.
[369] Ibid., 26.
[370] Weindling, *Health, Race and German Politics*, 179.
[371] Ibid.

ann's takeover of the publication *Archiv für Rassen- und Gesellschaftsbiologie* in 1918 marked a fundamental shift in the political orientation of Germany's racial hygiene movement.[372]

Lehmann was a respected publisher overall and frequently printed unbiased medical texts and journals. The publishing house did distribute medical journals and textbooks for students and doctors that were, on one level, strictly scientific. However, it was also subject on occasion to the racial agenda of its founder and served to "racialize" medical science.[373] The publishing house once boasted that it had never allowed Jews to work in any positions of power.[374] Consequently, Lehmann has rightfully been referred to as "the most powerful force for ideological unity among right-wing racist culture"[375] and as having assisted in forging an alliance between the political right and racial hygienists.[376] Lehmann did much to create a climate of opinion that blurred medical fact, Aryan supremacy, and German nationalism.

Lehmann's membership in the institute and the inclusion of a number of other strongly Nordic-minded individuals suggests that the German Institute for Blood Group Research appealed to a certain political faction. Eugenicists Eugen Fischer and Erwin Baur, Philalethes Kuhn, Hanno Konopacki-Konopath, and the architect Paul Schultze-Naumberg were all members. Philalethes Kuhn joined the National Socialist Party in 1923. Konopacki-Konopath, a Ministry of Reconstruction civil servant, acted as editor-in-chief of the far-right review *Die Sonne* (The Sun) and in 1926 founded the "Nordic Ring," a league of Aryan supremacist groups, of which fellow institute member Paul Schultze-Naumburg was a leader.[377] He also authored *Kunst und Rasse* (Art and Race) and had been an acquaintance of Hitler's since 1926.[378] Emphasis seemed to weigh heavily on one's political sympathies, and less on anthropological or medical expertise.

In addition to its focus upon native Germans and their membership profile, the Institute for Blood Group Research further demonstrated racial

---

[372] Proctor, *Racial Hygiene*, 27.
[373] Weindling, *Health, Race and German Politics*, 471.
[374] Proctor, *Racial Hygiene*, 26.
[375] Weindling, *Health, Race and German Politics*, 471.
[376] Proctor, *Racial Hygiene*, 27.
[377] Weindling, *Health, Race and German Politics*, 474.
[378] Ibid., 279.

bias by its methodology—namely, its consistent reliance upon physiognomy in racial categorization. Racial anthropologists routinely referred to visible traits in their research, but the conviction that perceptible differences existed among the racial types was most pronounced among *völkisch* race theorists. Under Reche's instruction, studies directed by the German Institute for Blood Group Research were to survey not only blood type, but also individuals' pigmentation and anatomical shapes—the "default" means of racial identification with which Reche and his colleagues would have been most familiar. On a more practical note, because the primary objective of the institute was to determine the relationship between race and blood, it was necessary to place seroanthropology in the context of existing methods of racial analysis. In 1926 the institute prepared a form giving researchers detailed instructions on which traits to record. In addition to such information as the subject's surname, mother's maiden name, and birthplace, columns were also provided for their hair and eye color, skull and face shape. [379] To make certain that examiners who had not received any "systematic training" in anthropological measurements would make no "significant mistakes," detailed explanations were provided.[380] (See Figures 7, 8, and 9.)

For those who wanted to see a connection between blood type and racial or genetic inferiority, it was there. *Völkisch* theorists, driven by their belief in biological racial differences, were more likely to have a positive outlook on seroanthropology and draw attention to correlations that had been made. The labeling of type B as an Eastern trait was confirmed by its relative infrequency in remote areas of Germany. There were also associations made between type B blood and non-German traits such as "foreign-sounding names, round skulls, darker pigmentation, and even a mongoloid appearance." There had also been findings to suggest that type B blood was potentially a marker of genetic inferiority. Max Gundel reported that frequencies of type B blood among the late-stage syphilitics he examined

---

[379] These are the criteria listed in Form 8d of the German Institute for Blood Group Research. For a blank version, see Steffan, ed., *Handbuch*, 390–391. There were also spaces for the subject's sex, age, first name, last name, place of residence, birthplace, parents' religion, and the birthplace of each parent. Hair color was recorded as white blond, yellow blond, dark blond, brown, black, blue black, red blond, red brown, or "fox" red (*Fuchsrot*, a very bright red). Eye color (only the color of the iris was examined) was classified as blue, gray, blue outer ring with orange inner ring, blue outer ring with brown inner ring, green, gold brown, brown, dark brown, or black. Ibid., 392.

[380] Ibid., 393.

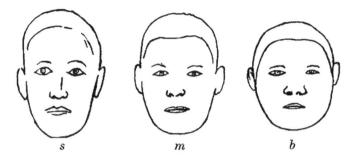

Figure 7. Diagram of facial shapes to aid examiners in shape classification, from the German Institute for Blood Group Research. From left to right: thin, medium, wide (schmal, mittelbreit, breit). Paul Steffan, "Die Bedeutung der Blutgruppen für die menschliche Rassenkunde," in Handbuch der Blutgruppenkunde, ed. Paul Steffan (Munich: J.F. Lehmann, 1932), 393.

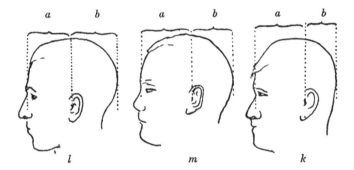

Figure 8. Profiles showing long, medium, and short skulls (left to right: lang, mittellang, kurz), from the German Institute for Blood Group Research. Paul Steffan, "Die Bedeutung der Blutgruppen für die menschliche Rassenkunde," in Handbuch der Blutgruppenkunde, ed. Paul Steffan (Munich: J.F. Lehmann, 1932), 392.

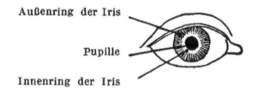

Figure 9. Diagram to help determine eye color. Note the detailing of the "outer" and "inner" rings of the iris; color classification was very specific. Paul Steffan, "Die Bedeutung der Blutgruppen für die menschliche Rassenkunde," in Handbuch der Blutgruppenkunde, ed. Paul Steffan (Munich: J.F. Lehmann, 1932), 392.

were more than double that of the general population, which led him to theorize that type B blood "uniquely predisposed" one to infection with the disease.[381] Previously, Gundel had noted elevated frequencies of type B blood among criminals. Dr. K. Boehmer analyzed a separate group of prisoners and reported the same pattern.[382] Boehmer found that his subjects' levels were twice that of the normal population, and when he, like Gundel, recalculated his results according to conviction and sentence, this disparity became even more pronounced. The percentage of type B blood was much higher among habitual criminals, those who had been given particularly lengthy sentences, or had been categorized as "incorrigible" (*unverbesserlich*).[383] Some researchers had further implied the inferiority of type B blood through even more obscure traits. In an article published in the *Münchener Medizinische Wochenschrift*, one physician claimed that individuals' "time of defecation" varied by serological type; on average, those with blood type A took only a few minutes, yet it took those with type B quite a bit longer—anywhere from twenty to forty minutes.[384] Another researcher claimed that blood type B predisposed one to infection with syphilis.[385]

At the same time, substantial evidence argued against any connection whatsoever between blood type and other characteristics—anthropological, pathological, or otherwise. Some studies suggested that type A blood, not B, was the "inferior" one. Based on the fact that there were more individuals with type B among his subjects over twenty-five years of age, a Danish contributor to the *Zeitschrift für Rassenphysiologie* initially suggested that individuals with type B blood lived longer than the other three types (although a later examination of 2,000 additional samples yielded "a completely different result").[386] Polish physician Eugeniusz Wilczkowski

---

[381] Max Gundel, "Bestehen Zusammenhänge zwischen Blutgruppe und Luesdisposition sowie zwischen Blutgruppe und Erfolg der Luestherapie?" *Klinische Wochenschrift* 6, no. 36 (September 1927): 1704.

[382] K. Boehmer, "Blutgruppen und Kriminalstatistik," *Forschungen und Fortschritte* 3 (1927): 141–142.

[383] Although his results do not list physiognomic characteristics, Boehmer noted the subjects' body types and commented on a tendency among those with type B towards an "asthenisch" (slender, or weak) body type.

[384] Type O individuals were referred to as "somewhere in between." Warnowsky also explained that, after an evening of "excessive alcohol intake," an individual with type A blood suffered the next day, but his cohort with type O felt even better and "more refreshed" than before. Drinkers with type B blood fell somewhere in between these extremes. See J. Warnowsky, "Über Beziehung der Blutgruppen zu Krankheiten: Heterohaemagglutination," *Münchner medizinischer Wochenschrift* 74 (1927): 1758–1760.

[385] Geisenhainer, *Rasse ist Schicksal*, 137.

[386] Sören Hansen, "Methodologisches über Blutgruppenforschung durch Massenuntersuchung," *Zeitschrift für Rassenphysiologie* 2 (1929/1930): 74.

reported that the schizophrenics he examined tended to have type A blood. In a survey of 500 subjects with "endogenous psychoses" (in this case, schizophrenia and manic depression), Ukrainian doctors B. Chominski and L. Schustova also found a relative predominance of type A.[387] Overall, the findings canceled out any perceptible relationship. German physicians C. Dolter and M. Heimann reported no difference between the blood type distributions of Germans with mental disorders and those without. The same was true in a study of inmates at an asylum in Silesia.[388]

Gundel's claim that type B individuals were more likely to contract syphilis was also disputed. After supplementing his own subject group of approximately 400 men with these figures, Gundel was available to expand his subject base to over 10,000 individuals.[389] He then "discovered" what many were already aware of: syphilitics were not more likely to have type B blood, or any other type for that matter. Faced with these results, Gundel was forced to revise his initial claim and admit that blood type appeared to play no role in susceptibility to syphilis.[390] Despite the wide range of

---

[387] Subjects were patients at the Ševčenko Hospital for the Mentally Ill in Kiev. See B. Chominskij and L. Schustowa, "Zur Frage des Zusammenhanges zwischen Blutgruppe und psychischer Erkrangung," 305. This was, however, a very brief notation. Their use of the term "relative" was quite liberal: with 44.3 percent A blood, the schizophrenics did have *slightly* more than the general populations' 43 percent. The difference was clearly too small to make any larger hypotheses.

[388] Fr. Meyer, "Die Blutgruppenverteilung in der schlesischen Bevölkerung sowie die Beziehung der Blutgruppen zu Geisteskrankheiten," *Deutscher Medizinischer Wochenschrift* 54, no. 35 (1928): 1461.

[389] Gundel, "Bestehen Zusammenhänge zwischen Blutgruppe und Luesdisposition sowie zwischen Blutgruppe und Erfolg der Luestherapie?" 1704.

[390] Even in the late 1920s, some physicians suspected that a particularly virulent infection could even change an individual's blood type. Records from the Robert Koch Serological Institute detail one such case reported by a Greek physician, Dr. Jack Diamontopoulos. Diamontopoulos claimed to have observed changes in the blood types of two patients. The first patient, suffering from gonorrhea, had arrived towards the end of April 1926 and, after receiving treatment, left the hospital on June 10, 1926. She returned again on January 1, 1928, with a secondary syphilis infection and, after ten injections of neosalvarsan [a mercury-based chemical used to treat syphilis], was cured of this as well. She was then released from the hospital a second time on March 31, 1928. According to Diamontopoulos, the patient had blood group B when tested on January 26, 1928. However, five weeks later, she was now type O!
The second patient came to the hospital in mid-August, 1926, also due to a gonorrheal infection. She was treated and left the hospital on December 31, 1926. She returned in August of the following year due to a primary syphilis infection. After having received extensive treatments (probably also salvarsan injections), she was released on October 27, 1927. In February 1928 she was readmitted to the hospital because of gonorrhea. A test in January 1928 indicated that she had blood type B; after four weeks, however, Diamontopoulos claimed that she had blood type O. After another four weeks, he noted, she "again proved to have blood type B." Diamontopoulos was convinced that the apparent change in blood type resulted from immunities built against the infections; this had enacted a change in the quantity and/or quality of the blood's agglutins. He also referred to the work of Raphalkes, who purportedly had also observed changes in blood type following extreme infections. The Robert Koch Institute simply stated that these cases described by Diamontopoulos "could not be considered [reliable] evidence against the constancy of the blood groups." His reference to

pathologies and disorders, including "schizophrenia, manic depression, progressive paralysis, epilepsy, imbecility, idiocy, senility, and dementia," no pattern could be noted between these and serological type.[391]

The same applied to antisocial or immoral behavior. Dr. Augustin Foerster, a physician in Münster, criticized both Gundel and Boehmer for likely overestimating their results in their studies of prisoners' blood types. [392] After compiling the statistics on his own subject of penitentiary inmates, also classed as the "most severe" type of criminal, he could not report an increase in type B blood. To ensure that his research was comparable to that of Gundel and Boehmer, Foerster made certain that his subjects were either habitual offenders, categorized as "incorrigible," or were the recipients of particularly lengthy sentences. Instead of being elevated, Foerster found, the level of type B blood among his subjects matched that of the general population.[393] In fact, it was even slightly lower among those individuals found guilty of the most violent crimes (rape, murder, manslaughter). In Gundel's research, the same class of convicts had nearly three times this amount of type B blood. Based on these figures, Foerster maintained that type B blood's association with criminal behavior was mistaken; its increased frequency among criminals "could not be confirmed."[394]

In addition, "Western European" type A blood did not reliably correspond with "Aryan features" or "superior" genetics—just as "Eastern" type B blood did not always mean dark features, round skulls, or various mentally or physically debilitating conditions. (See Figure 10.) After his analysis of soldiers in East Prussia, Reich found himself puzzled by the *low* incidence of type B blood among subjects with both Eastern surnames and

---

"frequently occurring pseudoagglutinations," which the Koch Institute believed to be quite rare, led them to speculate that perhaps Diamontopoulos had used "defective techniques" in his analysis. Diamontopoulos had failed to indicate in his work that he had analyzed not only the blood types, but also the serum of the concerned patients. Therefore, the Koch Institute judged his data on both cases to be "incomplete." In the past, they emphasized, all researchers who had followed the correct procedure had consistently observed the constancy of the blood groups—even after cases of extreme sickness and/or particularly aggressive medical treatment.See R86/3781.

[391] A. Foerster, "Blutgruppen und Verbrecher," *Deutsche Zeitschrift Gerichtliche Medizin* 11, no. 6 (1928): 488.

[392] Ibid. Foerster was at the Institute for Forensic and Social Medicine in Münster.

[393] Of the 371 individuals Foerster examined, 136 men fit into one of these categories: 41.3 were type A, and only 12.7 percent type B blood. Type O was 41.7 percent and AB was 4.3 percent.See Ibid., 489.

[394] In addition, Foerster had not found that type B individuals were predisposed to one specific body type, whereas Boehmer noted that type B persons were inclined to be "athletic" (Gundel had not considered body types). Like Fritz Schiff, Foerster was rightly wary of any studies that drew correlations between physiognomic characteristics and social status. See Schiff, "Zur Serologie der Berliner Bevölkerung," 448–450.

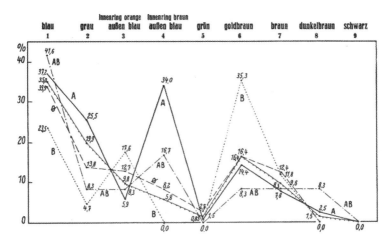

Figure 10. Searching for correlations: A detailed graph by Paul Steffan comparing the blood types and eye colors of 256 subjects. Paul Steffan, "Die Beziehung zwischen Blutgruppe, Pigment und Kopfform," Zeitschrift für Rassenphysiologie 1 (1928/1929): 138.

"Mongoloid" eye shapes. Klein, too, unexpectedly found "contradictory results" among his Rhineland schoolchildren linking type B blood to blond hair and type A to dark hair. Like Reich, he attributed this inconsistency to racial mixing. Presumably, had these subjects not been products of miscegenation, their serological type and physiognomy would have matched.[395] Klein claimed that any conclusion was "improbable on account of the complexity [of the results]."[396] In another study on Rhinelanders, which set out specifically to determine whether a connection existed "between blood type and physiognomy," the authors found no recognizable pattern.[397] To complicate matters further, there were wide discrepancies in blood type distributions even within Germany. In 1928 the highest incidence of type A within a population (76.4 percent) was recorded in Nordostharz; the lowest (21 percent) was in Bavaria. Type B was most common in Westphalia, at 23.4 percent, but this dipped to only 6.4 percent in Baden.[398] Inexplicably high reports of type B blood were not uncommon; it was nearly 18 percent in Giessen and Vogelsberg, and this increased to 20 percent

---

[395] Reich, "Soldatenuntersuchungen nach Blutgruppen und Konstitutionstypen," 149.

[396] Klein, "Ergebnis der Blutgruppenbestimmung in Oberlahnstein und St. Goarshausen," 12.

[397] T. Wohlfeil and F. Isbruch, "IV. Mitteilung Über die Blutgruppenverteilung im Rheinland. V. Mitteilung Zur Frage der Korrelation Zwischen Blutgruppe und Anthropologie," Klinische Wochenschrift 8, no. 47 (1929): 2186.

[398] Ibid.

in Wetterau, and then 25 percent in northern Wetterau.[399] There had also been statistics that researchers were simply not sure how to interpret. Near Frankfurt am Main, researchers were bewildered by an unusual frequency of type AB blood that was not found elsewhere in Frankfurt, or even in Germany.[400] Schaede was "puzzled" by the high frequencies of O and AB in Reich's study. Schridde, also, mentioned the interesting high incidence of type O blood among his subjects.

Outside Germany other researchers, too, had commented on discrepancies between blood type and appearance. Portuguese physician Antonio Augusto Mendes-Corrèa noted that peoples of different serological types could resemble one another. The reverse might also be true —populations that shared the same distributions of blood type could look very different. This gave Mendes-Corrèa "very strong reservations" (*une réserve très forte*) about the taxonomic value of the blood types.[401] American Ella Grove had also reported that the blood type distributions of the Ainu of Japan, supposedly a Caucasian people, were found to be the same as "Senegal Negroes." The two looked completely different, and were regarded as separate racial types by anthropologists, yet had similar blood type distributions. The findings suggested to Grove that blood was useless as a racial indicator. British physician Matt Young also observed how even peoples far removed from one another and "divergent in physical type" could have identical blood type distributions, as was the case between the Senegalese and Sumatran peoples. Such discrepancies similarly led him to question altogether the value of blood typing for racial differentiation.[402] In 1927 one physician dryly commented, "as far as racial studies are concerned, we still do not know whether or not the blood types are useful."[403] For the purposes of racial classification, most racial analysts still preferred appearance. In spite of the latter's "imperfections and faults," Portuguese physician Mendes-Corrèa believed that it was still more efficient than blood science.[404] Matt Young similarly cautioned against relying upon the existing

---

[399] Foerster, "Blutgruppen und Verbrecher," 489–490. See also H. Kliewe and R. Nagel, "Über die Blutgruppenzusammensatzung der Bevölkerung Oberhessens," *Klinischer Wochenschrift* 6, no. 49 (1927): 2332–2335.

[400] Wohlfeil and Isbruch, "Zur Frage der Korrelation zwischen Blutgruppe und Anthropologie ," 852.

[401] Mendes-Corrèa, "Sur les prétendues 'races' sérologiques," 439.

[402] M. Young, "The Problem of the Racial Significance of the Blood Groups," *Man* 28, no. 9 (1928): 171.

[403] Mendes-Corrèa, "Sur les prétendues 'races' sérologiques," 332.

[404] Ibid., 322.

evidence, as it did not seem to be enough to attach more importance to blood in racial analyses than other racial traits—such as "head or hair form or skin color."[405] The once-committed Polish blood scientist Jan Mydlarski, previously a student of Ludwik Hirszfeld, now believed that blood science had only a "secondary value" in anthropological studies.[406]

In spite of increasing criticism directed towards seroanthropology, and the discipline's frustrating intricacies, there were other physicians, and indeed some of the same who were perplexed with their own findings, who nevertheless advocated further research before completely abandoning racial studies of blood. Ella Grove believed that even more studies might better explain the nature of the relationship between blood type and appearance. She suggested continued blood type surveys with other characteristics, such as "the color of the eyes, hair, and skin," as this might "provide insight on the complete [racial] identity of the individual."[407] Her colleague, Dr. Clark Wissler, agreed; the only way to prove the value of the blood test was "to take some definite anatomical character and correlate it with blood type."[408] Karl Landsteiner proposed further analysis of blood in the context of other racial traits. It was not possible, he pointed out, to classify races by measuring just the nose or any other one trait, and this applied to blood as well—blood type was simply "one factor" in (racial) classification.[409] Further studies would also explain the significance of the other, "neglected" blood types O and AB, the racial value of which was often discounted on the grounds that both were relatively recent developments in human history. Many researchers theorized that they were probably the result of mutations after the racial types had emerged. Therefore, only A and B could be used to study anthropology. Some, however, argued that O and AB were not the product of mutation, but were actually much older than originally thought, and therefore just as important for racial research as types A and B.[410]

---

[405] Young, "The Problem of the Racial Significance of the Blood Groups," 172.

[406] A.A. Mendes-Corrèa, "Sur la valeur anthropologique des groupes sanguins," *Sang* 1, no. 4 (1927): 323.

[407] Grove and Coca, "On the Value of the Blood Group 'Feature' in the Study of Race Relationships," 91.

[408] Ibid.

[409] Ibid.

[410] In 1925 Russian physician Wischnewsky introduced a racial-serological index that took into account all four blood types. See B.M. Wischnewsky, "Zur Frage der biochemischen Rassenindex," *Wratsch. Djelo* 1, no. 6 (1925): 484, and "Blutgruppen und Anthropologie," *Verhandlungen. der Ständigen Kommission für Blutgruppenforschung* I, no. 2 (1927): 1–25. See also U.U. Melkich, "Der neue biochemische Rassenindex," *Verhandlungen der Ständigen Kommission für Blutgruppenforschung* 1, no. 2 (1927): 26–39.

Others pointed to the haphazard development of this new science as a potential source of inconsistencies. Studies of blood and race had been erratic; general guidelines had yet to be established. The size of the subject group was important, as making any type of definitive statement first required that a sufficient number of individuals be examined, though the number necessary was subject to debate. In his study of 660 soldiers, Reich added that his results could not be applied "without reservation," as his subjects were only a small sample of East Prussia's population—though he failed to specify what percentage of the population would, ideally, need to be typed. At the same time, in Klein's analysis of approximately 600 schoolchildren, he did not refer to any such requirements. Even significantly larger studies, on occasion, were criticized for insufficient sampling. Matt Young expressed strong reservations about Verzár and Wesczeky's 1921 study, which concluded that, based on the comparable distributions of blood type between Hungarian Roma-Sinti and Indians (drawn from the Hirszfelds' 1918 research), blood type was an unchanging racial characteristic. Young claimed that neither study had tested enough people native to the region to arrive at such a conclusion:

> It is very doubtful if the results of 1,000 observations can be accepted as in any way representative of the blood grouping in India, in which there appears to be evidence of several racial types. Perhaps too much emphasis has been laid on this particular example of apparent persistence of blood type.[411]

Otto Reche, intent on justifying his own racial convictions through medicine, predictably downplayed discrepancies in seroanthropology and simply claimed that further research was required.[412] In 1928 he responded to misgivings in a *Volk und Rasse* article entitled "Blood Group Research and Anthropology." Reche theorized that blood types A and B had, at one point, been connected with anthropomorphic characteristics; the lack of correlation between the two in modern analyses, he said, was a result of miscegenation. He explained how this same phenomenon could be observed in the physical characteristics of "racially mixed" individuals (*Mischlinge*) who, with their light hair, skin, and eyes, had an "Aryan appearance" that belied

---

[411] Young, "The Problem of the Racial Significance of the Blood Groups," 172.

[412] Geisenhainer, *Rasse ist Schicksal*, 130. These developments might have played a role in the increasing cancellation of memberships in the institute, though many cited financial reasons for doing so.

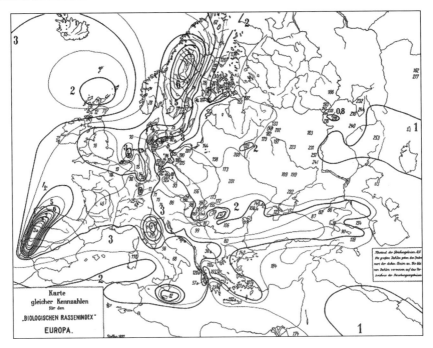

Figure 11. Map by Paul Steffan showing Hirszfeld's biochemical racial indices
throughout Europe as of 1927. Paul Steffan and Siegmund Wellisch,
"Die geographische Verteilung der Blutgruppen," Zeitschrift für Rassenphysiologie 1 (1928–1929): 46–60.

their true racial state. According to Reche, sorting the traits of blood and
physiognomy only required further research. There was the additional possibility, he mused, that it might not be physiognomic traits, but psychological ones, that went "hand in hand" with the blood types.[413] Reche made
fleeting reference to promising new research which suggested that a relationship might exist. To him, seroanthropology still seemed a worthwhile
pursuit. (See Figure 11.) He implored blood scientists to enthusiastically
continue their work.[414] Reche's commitment to seroanthropology sets him
apart from a general increasing indifference to the discipline.

Reche's conviction that race was a definitive biological category
remained, even in the face of adversity. In 1921 Reche first described race
as a number of distinct, consistently occurring "morphological, phys-

---

[413] Otto Reche, "Blutgruppenforschung und Anthropologie," Volk und Rasse 3 (1928): 8.
[414] Ibid., 10–11.

iological, and psychological hereditary traits." He added that one race could be sharply separated from the next by its *"Volk*, language, and cultural community."[415] Nearly a decade later, after he co-founded the German Institute for Blood Group Research, his opinion of what defined "race" had changed little:

> A race is a group of people connected by a blood relation...they demonstrate this relation through shared hereditary dispositions as well as certain characteristics...Racial characteristics consist of not only externally visible physical...features, but also of physiological and spiritual traits...and just as all "culture" and "civilization" are products of both hereditary racial predispositions and "environment," the vast differences in the sophistication and the manifestation of cultures and civilizations lead us back, for the most part, to the different spiritual dispositions of the races; it has been repeatedly observed that a culturally creative race (like the Northern Europeans) can produce cultural value in an unfavorable environment, while an untalented race under favorable conditions doesn't rise out of its primitive state. The most important visible characteristics are as follows: the color of the skin, eyes and hair, the texture of the hair, the shape of the skull and face, the shape and proportion of the limbs and torso, and general body measurements.[416]

*Völkisch* studies of blood and race reflected this same ambiguity. Klein, for instance, remarked on "a certain measure of Spanish blood" in the Germans he examined, but he did not refer to specific blood type proportions or any other clinical details. To Reich, type B blood was the Eastern type, but he failed to explain which racial types were "Eastern." Other, similarly vague phrases were used; Schaede took pride in his "regularly mixed" population, as did Schridde in his "unmixed" one. Steffan declared that there had been no "Jewish influence" in Osnabrück, and Schridde claimed that his forest settlements were "untouched." Such wording allowed for a great deal of subjectivity, so its usage was common among those who had no definitive proof of race, but maintained it existed nonetheless.

After several years and dozens of further blood type surveys, seroanthropology faced similar circumstances as it had when the German Insti-

---

[415] Otto Reche, "Rasse und Sprache," *Archiv für Anthropologie* 46 (1921): 208.
[416] Quoted in Geisenhainer, *Rasse ist Schicksal*, 139–140.

tute for Blood Group was first established. Some were still intrigued by the potential of the science, but most quickly disregarded it as unreliable. The *völkisch* blood scientists of the institute expected research to reveal an association between type B blood and Eastern peoples, as well as greater susceptibilities to mental and physical illness or disability. Membership in the German Institute for Blood Group Research weighed heavily on one's political inclinations and, possibly, as the lists indicate, their "racial type"; the impressive but relatively liberal contingent of Jewish blood scientists was largely absent. Institutional research interests were biased as well— desperate attempts were made to draw some correlation between blood and other characteristics. By the late 1920s, however, the tangled consistencies and inconsistencies of blood type surveys led to serious misgivings concerning the anthropological significance of blood type.

# CHAPTER V

# THE JEW AS EXAMINER AND EXAMINED

Jews, or individuals of "Jewish descent," figured prominently in both the development and advancement of blood science. Karl Landsteiner launched the discipline of serology with his discovery of the blood types in 1900. In 1906 German bacteriologist August von Wassermann developed a blood test for the diagnosis of syphilis that allowed for early detection of the disease and became the standard means for syphilis testing for much of the twentieth century. In 1918 Ludwik Hirszfeld was the first to trace a link between blood type and race. In 1924 in Berlin, the theory of blood group tests as a method to prove non-paternity was introduced by Fritz Schiff for the first time. Germany, followed by other European countries and then the United States, came to permit blood tests for paternity testing because of his work.[417] In that same year, Felix Bernstein worked out the correct hypothesis for the genetic transmission of the blood types.[418] In spite of these contributions, Jewish membership in the Institute for Blood Group Research was conspicuously lacking.

No reference as such was made to the racial inferiority of the Jews in the *Zeitschrift für Rassenphysiologie*, the publication of the German Institute for Blood Group Research, but this was implied through studies of racially pure (Aryan) Germans to the exclusion of Jews. Furthermore, Reche and Steffan's anti-Semitism was plain in their research—witness Steffan's reference to "no Jews" in an isolated area of Germany or Reche's observation

[417] M. Okroi and P. Voswinckel, "'Obviously Impossible'—The Application of the Inheritance of Blood Groups as a Forensic Method. The Beginning of Paternity Tests in Germany, Europe and the USA," *International Congress Series* 1239 (2003): 711–714.

[418] Norbert Schappacher, "Felix Bernstein," *International Statistical Review* 73 (2007): 3–7.

that his "inferior" South Pacific subjects "looked Jewish." The fact that Jewish blood scientists predominated in their field, yet were noticeably absent from the German Institute for Blood Group Research highlights further the fact that the institute attracted a certain political contingent. The Jews' supposed racial otherness and their intermarriage with other Germans were matters of concern to *völkisch* race theorists. For this reason, Jews were also a prioritized subject group in studies of blood and race, with the expectation that they were a race apart from the majority Christian population. Jews had a unique relationship to seroanthropology, however, because at the same time, Jewish blood scientists also carried out racial analyses of blood. This resulted in a strange tension between Jews' position as researchers and also as subjects. Studies of Jews' blood types were concurrent with the Nazis' rise to power; however, such studies did not elicit the interest of party members. Nonetheless, it is important to take seroanthropology into consideration because it constituted a branch of racial science—and racial difference was central to the Nazi worldview. Differences in "blood" and "blood defilement," simple regurgitations of the racial blood rhetoric so commonly used in the pre-World War I and Weimar eras, would continue to be referenced in National Socialist propaganda.

Seroanthropology seemed to have all the necessary components of a science that the Nazis would have used to their advantage. Driven by the belief that racial classification of the Jewish minority would benefit the well-being of the "Aryan people," German anthropologists consistently studied Jews to rationalize their own anti-Semitism and to justify Jewish exclusion from the *Volksgemeinschaft*, or "national community." Anti-Semites often claimed that Jewish ancestry was visible and made reference to a lengthy list of "typically Jewish" traits as proof: dark pigmentation, disproportionately large noses, and small, closely set eyes. Still others cited excessive body hair, narrow chests, or lack of musculature. However, anthropologists could not deny that these traits were quite often lacking amongst Jews, and that there were many Jews who "looked Nordic." This left the uncomfortable question of how race manifested itself, if not in appearance. If this were the case, it would make physical indicators useless. For some, there was the lingering possibility that the Jews were "partly Nordic." American anthropologist Stephen Rich pointed out that hair, eye, and skin color were strictly hereditary and not subject to environmental change. Consequently, he observed, if the Jews were direct, pure descendants of the ancient "Hebraic racial type," they would not have blond hair, blue eyes, or the typically Teutonic fair complexion that many of them did—as none of these characteristics were found among their darker presumed congeners.[419] "Certainly," Rich observed, "there were more than one-quarter blue-eyed or rufous (red-haired) Jews among those from Germany and the Ukraine."[420] Anthropologist Rudolf Virchow similarly reported that a substantial percentage of Jewish children possessed these traits. Several studies in physical anthropologist Maurice Fishberg's book indicated 30 percent blond hair among Jews, which led to one estimation that the Jews had at one point had at least 50 percent blond ancestry (as blondness is a recessive Mendelian trait).[421] Frustrated by these "inconsistencies," many anti-Semitic anthropologists claimed that a psychological assessment was also necessary in the racial classification of Jews, as their race was uniquely predisposed to immorality and mental illness.[422]

Matters were further complicated by conflicting claims as to the Jews' racial lineage and classification. Anthropologists commonly divided Jews into the Sephardi and Ashkenazi racial types. Sephardim were described as dark with "long skulls" and Ashkenazim as fair with "round skulls." Others expanded this categorization. German anthropologist Felix von Luschan theorized that modern Jews had descended from three groups: Semites, Aryan Amorites and, predominantly, descendants of the ancient Hittites. Nineteenth-century race propagandist and fervent German nationalist Houston S. Chamberlain also believed that the Jews represented a mixture of Semitic, Indo-European, and Hittite peoples. Austrian Jewish anthropologist Ignaz Zollschan claimed that Jews were the result of "miscegenation between two ancient peoples"—the Egyptians of North Africa and the Mesopotamians of the Near East—which made them a mix of "mainly Oriental and Near Eastern" races. This hypothesis was popular among German

---

[419] Stephen G. Rich, "The Jews: Race or Conglomerate," *Journal of Educational Sociology* 2, no. 8 (1929): 474

[420] Ibid.

[421] Ibid. See also Maurice Fishberg, *The Jews: A Study of Race and Environment* (Newcastle-upon-Tyne: Walter Scott Publishing Co., 1911), 508.

[422] To substantiate this claim, many medical researchers made reference to the greater incidence of suicide among Jews. See, for example, statistics compiled on suicide in Prussia from 1912 to 1919, which revealed that approximately 2 percent of the suicides were Jewish. See Edgar Schultz, "Judentum und Degeneration," *Ziel und Weg* 16 (1935): 354.

race theorists: Eugen Fischer, Fritz Lenz, Hans F.K. Günther, and Ludwik F. Clauss all came to the same conclusion.[423]

When seroanthropology arrived, consequently, there was no general consensus as to either what "racial type" the Jews were, or how to go about differentiating them from non-Jews, in spite of their perceived threat to the racial integrity of the German *Volk*. Once it was realized that blood might have some relation to anthropology, this theory was immediately applied to the matter of the Jews' race. Jews had been one of the sixteen "racial types" analyzed by the Hirszfelds in their wartime study in Salonika. Blood type surveys of Jews were relatively common in the interwar years, particularly throughout Central and Eastern Europe. The surveys did not, however, provide any new insight into the matter of the Jews' racial identity. Based on their type distributions, the Hirszfelds had classified their Jewish subjects from Tunisia as Mediterranean, or "intermediate" in racial type— a category in between the "African-Asian" and "Western European" types. Many subsequent studies in Romania, Ukraine, Poland, and elsewhere in the Jewish Diaspora also classified the Jews as "intermediate" based on their blood.[424] Results were not consistent, however. Numerous analyses of Russian Jews categorized them instead as "Western European," and unexpectedly high levels of blood type A, comparable to or even higher than Germans examined, were reported among Jews from White Russia and Odessa.[425] Similarly, Fritz Schiff noted high levels of type A blood in a study of Berlin Jews, and the same pattern was observed in Holland.[426] Elevated

---

[423] Siegmund Wellisch, "Serologische Untersuchungen über das Rassentum der Juden," *Zeitschrift für Rassenphysiologie* 1 (1928/1929): 204.

[424] Romanian scientist Sabin Manuilă reported his 211 Romanian Jewish subjects to have an index of 1.54. Steffan, *Handbuch*, 408. Similarly, an analysis of 384 Ukrainian Jews by Russian scientists came to 1.63. Ibid., 410. A study published by Polish physicians W. Halber and Jan Mydlarski also classed 818 Jews as "Mediterranean" with an index of 1.94, but results were not always within this range. Ibid., 406.

[425] A survey of 257 White Russian Jews yielded a biochemical index of 2.42. A study of 1,475 Jewish subjects from Odessa gave a relatively high 2.20, while a separate survey of 529 Jews in the same region came to 2.96.

[426] Similar results were reported farther west. Schiff and Ziegler found their Berlin Jewish subjects to have an index of 2.4. Similarly, a result of 2.45 had been reported in a study of 705 Dutch Jews. For instance, see the following studies of Jewish subjects: Ukraine, 384 subjects, 1.63 biochemical index. Rubaschkin and Derman, "Die Hämoisoagglutination als Grundlage der Methode zur Erforschung der Konstitution," *Vrač. Delo*, nos. 20–23 (1924): 1153–1162. Poland, 818 subjects, 1.94 biochemical index. W. Halber and J. Mydlarski, "Recherches séroanthropologiques en Pologne," *Comptes Rendus de Société Biologique* 89 (1923): 1373–1375; White Russia, 257 subjects, 2.42 biochemical racial index. Rachowsky and Suetin, "Über eine Untersuchung von weissrussischen Juden," *Sapiski Bolr. San.-Bakt. Ist.*, no. 2138 (1926); City Odessa, 1,475 subjects, 2.20 biochemical index. A.L. Barinstein, "Zur Frage des biochemischen Rassenindex der Bevölkerung von Odessa," *Ukrainisches Zentralblatt für Blutgruppenforschung* 2 (1928): 55–61. Settlers near Odessa, 529 subjects,

frequencies of type A blood, the type also affiliated with "Aryan descent," were by no means uncommon among Jewish subject groups.[427]

Hoping to substantiate biological differences between the blood of Jews and non-Jews, *völkisch* blood scientists focused upon select differences, however slight, as evidence of the Jews' "racial otherness." Since references to the racial import of blood were so commonplace in the rhetoric of extremist groups, confirmation of an objective difference would have been practical from an anthropological standpoint, as well as politically expedient. The idea that intercourse with a Jew resulted in literal "defilement" of the German partner's blood, which formed the basis for the storyline in Artur Dinter's *The Sin against the Blood*, was a favorite theme in National Socialism. This was expressed in a poem simply titled "Blood" in a 1923 issue of *Der Stürmer*:

> As the blood is, so is the soul
> Women who let Jews have their way with them defile their body and soul.
> Whosoever carries Jewish blood in their veins thinks and behaves Jewish.[428]

More often than not, such propaganda was clearly allegorical, though it did on occasion assume medical objectivity. In a speech entitled "The People's Health through Blood and Soil," given in 1926, Streicher explained:

> It is established for all eternity; alien albumen is the sperm of a man of alien race. The male sperm in cohabitation is partially or completely absorbed by the female, and thus enters her bloodstream. One single cohabitation of a Jew with an Aryan woman is sufficient to poison her blood forever. Together with

---

2.96 biochemical racial index. M.S. Leitschik, "Die Blutgruppen und die Untersuchung der Vererbung der agglutinerendend Substanzen bie den übergesiedelten Hebräern der Odessaer Gebiete," *Ukrainisches Zentralblatt für Blutgruppenforschung* 2, no. 3 (1928): 30–44. Holland, 705 subjects, 2.45 biochemical index. M.A. van Herwerden, "Les groupes sanguins en Hollande," *Actes du congrés l'Institut International d'Anthropologie. III^{me} Session, Amsterdam* (1928): 456–461.

[427] See A. Eisenberg, "Zur Frage nach den Isoagglutinationsgruppen des Blutes bei Menschen," *Folia Haematologica* 36 (1928): 316–336. Of the 1,000 individuals examined from Charkow; 823 were Russian, 114 were Jewish, and the remainder represented different populations. The blood type distributions of the 823 Russians were as follows: type O 33 percent, type A 36.8 percent, type B 22.1 percent, type AB 8 percent, unclear .1. The distributions of the 114 Jews were: type O 27.2 percent, type A 44.8 percent, type B 21.9 percent, type AB 6.1 percent. The Hirszfeld Index was 1.81 for the Jews, and 1.53 for the group as a whole. That of the Russians was calculated to be 1.49.

[428] "Blut," *Der Stürmer* 1, no. 8 (July 1923): 2.

the alien albumen she has absorbed the alien soul. Never again will she be able to bear purely Aryan children, even when married to an Aryan. They will all be bastards, with a dual soul and a body of a mixed breed. Their children will also be crossbreeds; that means, ugly people of unsteady character and with a tendency to illnesses. Now we know why the Jew uses every artifice of seduction in order to ravish German girls at as early an age as possible; why the Jewish doctor rapes his patients while they are under anaesthetic. He wants the German girl and the German woman to absorb the alien sperm of the Jew. She is never again to bear German children. But the blood products of all animals right down to the bacteria like the serum, lymph, extracts from internal organs etc., are all alien albumen. They have a poisonous effect if directly introduced into the blood stream either by vaccination or by injection. By these products of sick animals the blood is ravished, the Aryan is impregnated with an alien species. The author and abettor of such action is the Jew. He has been aware of the secrets of the race question for centuries, and therefore plans systematically the annihilation of the nations which are superior to him. Science and authorities are his instruments for the enforcing of pseudo-science and the concealment of truth.[429]

In yet another issue of *Der Stürmer*, an excerpt told readers of blood group research that had confirmed that the blood of Jews "was entirely different from that of all non-Jewish races." A later report explained that Jews were a mix of the "German, Mongol, and Negro" races—and samples had revealed their blood to be "partly animal and ape" blood."[430]

Berlin doctor Fritz Schiff, a German Jew and one of the most respected serologists in interwar Europe, was quick to recognize the racist and political misappropriations of seroanthropology. Schiff and a colleague chose to investigate claims of uniquely "Jewish blood" by examining the blood types of German and German Jewish hospital inmates. The results indicated that the blood type distributions of Jewish subjects were comparable to those of the majority population. Schiff denied that blood enabled proof of (racial) difference but acknowledged that it might be useful "in assessing miscegenation." Clearly, his intent was not to completely discredit seroanthropology but to prevent its misuse. Explicit reference was made to the work of

[429] Quoted in Robert E. Conot, *Justice at Nuremberg* (New York: Carroll and Graf Publishers, 1984), 382.
[430] Ibid., 42.

Verzár and Weszeczky who, based on their 1921 study of Hungarians, settler Germans, and "Gypsies" in the region of Budapest, claimed that blood
type was a reliable racial indicator. Schiff dryly remarked on how their conclusions were based on small subject groups and how the study was conducted mainly by white researchers.[431]

Schiff's research and other studies that highlighted the similarities of
Jewish blood and German blood were largely ignored by seroanthropologists with far-right political sympathies. Such critics, eager to discredit
the possibility that there was no difference between German and Jewish
blood, drew attention to potential flaws in Schiff's research. Schiff had not
used a sufficiently diverse group, they claimed, and as the subject pool was
composed entirely of patients, they theorized that a "weaker disposition"
may have distorted the outcome. Furthermore, they had not compared the
subjects' physiognomies with their blood type—interestingly, this was an
intentional omission by Schiff and his colleague, who rightfully suspected
ulterior motives behind efforts to correlate visible "similarities or dissimi-
larities" with various groups of people. They cautioned researchers to be
more thorough in their studies before making any hypotheses. Because of
its larger implications that "Jewish blood" was no different than any other,
Schiff's work in Berlin was often referenced by impartial observers.

In the periodicals *Zeitschrift für Ethnologie* and *Morgen*, German Jewish
anthropologists Franz Weidenreich and Hans Friedenthalin reiterated the
study's main point that the blood group distributions of the Jews approximated those of their "host nation" and, as a result, no foreign blood as
such could be established.[432] The findings were also referred to in an article
entitled "The so-called human blood types and their incidence among the
Jews," published in *Jüdische Familienforschung* (Jewish Family Research),
which further emphasized the similarities between the two groups.[433] Schiff
attributed the Jews' comparable blood type distributions to their centu-

---

[431] Because the size of an "adequate" subject group was not standardized, Schiff and Ziegler did not elaborate on
how many individuals of each racial type were necessary (to make a hypothesis). This omission, and the simple fact that their own blood type survey in Berlin had even fewer subjects than that of Verzár and Weszeczky's, weakened Schiff and Ziegler's criticism.

[432] Franz Weidenreich, "Das probleme der jüdischen Rasse," *Morgen* 7 (1931): 78–96. See also Hans Friedenthalin, "Ober den Grad der Blutverwandtschaft in der Familie oder Sippenschaft," *Zeitschrift für Ethnologie* 40,
no. 8 (1916): 25.

[433] Fritz Schiff, "Die sogennanten Blutgruppen des Menschen und ihr Vorkommen bei den Juden," *Jüdische Familienforschung* 2, no. 4 (1926): 178–180.

ries-long existence in the Diaspora, as well as more recent assimilations. Of course, this implied that, at one point, the Jews did have a distinct blood type distribution. Simply put, their blood differed now because they had mixed with the Germans. Nevertheless, researchers stubbornly continued their efforts to discern Jewish from non-Jewish blood.

## E.O. Manoiloff's "Serochemistry" and Jewish Blood

In 1925 the German medical periodical *Biochemische Zeitschrift* ran the article "Attempts to Differentiate the Human Races through Blood." Its author, E.O. Manoiloff, claimed to have devised a technique that allowed the blood of the various races to be recognized through an intricate chemical process. Like many other race theorists, Manoiloff was initially drawn to blood as an alternative racial indicator to physiognomy. He found the popular skull-index particularly frustrating. "Skulls of the same size," Manoiloff complained, "belong to the most different nationalities with the most different civilizations."[434] By contrast, he declared that his new serochemical process—the result of "years of research"—was much more reliable.

Unlike previous studies, Manoiloff did not simply type his subjects' blood and then analyze the distribution patterns. Instead, he modeled his approach on hormonal differences in blood. Scientists were aware of the presence of hormones in blood, and these could be used to determine whether a sample had been taken from a man or a woman. Manoiloff reasoned that this same principle might apply to race. He conducted chemical analyses of blood samples from different races, but mainly Russians and Russian Jews, selected on account of the "many circumstances" that ensured that they would be completely different from one another. Manoiloff theorized that miscegenation had been relatively uncommon among Russian Jews.[435] To further diminish this risk, however, the only individuals examined were those with at least three generations of "exclusively Russian or Jewish ancestry" on both their maternal and paternal sides.[436]

---

[434] E.O. Manoiloff, "Discernment of Human Races by Blood: Particularly of Russians from Jews," *American Journal of Physical Anthropology* 10, no. 1 (1927): 12.

[435] Ibid., 16.

[436] Anna T. Poliakowa, "Manoiloff's 'Race' Reaction and Its Application to the Determination of Paternity," *American Journal of Physical Anthropology* 10, no. 1 (April 1927): 24.

Manoiloff's technique was unique among seroanthropological studies in that blood type was irrelevant, and not even necessary, for his purposes. Instead, five separate chemical reagents were used that reacted with
the blood sample to indicate the individual's race. If administered correctly, Manoiloff explained, the mixture would turn a characteristic color.
For instance, the presence of Jewish blood in the test tube would cause the
sample to take on a greenish cast. Russian blood could also turn a shade of
green, but it was a "much darker green," and the two were further differentiated by reaction time, which was reportedly "much quicker" with Jewish
blood. After testing this reaction on 2,000 Russians, 800 of whom were Jewish, Manoiloff claimed that he had been able to distinguish the blood sample of a Jew from that of a Russian 88.6 percent of the time. Encouraged by
these results, and in anticipation of presenting his findings at the upcom-
ing Eighth Conference of Russian Therapists, he solicited the assistance of
colleagues in collecting additional blood samples. In his request, Manoiloff reiterated that the sample needed to come from an individual with an
appropriately "pure" racial lineage. To make it a blind study, and thereby
better determine the consistency of his method, he further requested that
contributors "mark the test tubes only with numbers, without either family
names or designation of nation (nationality)."[437] Though several physicians
responded, the number of submissions was minimal.[438]

The purported accuracy of the technique did increase with the somewhat larger subject pool. Manoiloff now claimed that he was able to correctly identify the subject's race 91.7 percent of the time.[439] To race theorists, the prospect of identifying a subject's race from a blood sample alone
was intriguing. Taken at face value, the simplicity and reliability of this technique were much more impressive than either conventional physiognomic
methods of racial research or surveys of blood and race. Seroanthropological research focused only on the distributions of blood types within a population; from a racial perspective, an individual's blood type was unimport-
ant outside of the context of the group to which he or she belonged. Initially,

---

[437] Manoiloff, "Discernment of Human Races by Blood: Particularly of Russians from Jews," 16.

[438] Their subject groups were often small. Professor P.S. Medowikoff provided samples from thirty-seven Jews
and forty-four Russians. Dr. Rachel Liwschitz supplied samples from sixteen Jews and twelve Russians. P.S.
Medowikoff and Rachel Liwschitz, physicians at the German Red Cross Hospital, and others throughout
Russia submitted samples.

[439] Manoiloff, "Discernment of Human Races by Blood: Particularly of Russians from Jews," 17.

Manoiloff's technique seemed to offer an objective means of racial classification. It also seemed to have potential with other racial types; he reported
that his method also enabled him to distinguish among "Chinese, German,
Estonian, Finnish, Polish, and Armenian blood," though the subject groups
had admittedly not been large enough to make any substantive claims.

In spite of its scientific overtones, Manoiloff's technique was far from
objective. Due to the similarity in reaction colors, classification at times had
to rely upon slight differences in shades, which made this, too, a subjective
process. In addition, the obscure chemical reaction met mainly with criticism and doubt from the medical community. This skepticism was certainly
made worse by Manoiloff's response. When questioned about the specifics
of the reaction, he admitted that he could not say much, mainly because
the part played by the critical component cresyl-violet was unknown.[440] As
a result, Manoiloff's technique attracted only minor attention. Even though
the statistics were not nearly as precise, studies of blood and race continued
to rely upon the usual survey methodology.

## SEROANTHROPOLOGICAL ANALYSIS OF JEWS

The German Institute for Blood Group Research promptly addressed the
serological identity of Jews; the introductory issue of the organization's
periodical included an article on the topic. Its author, Siegmund Wellisch,
an engineer employed with the Vienna Magistrate, believed that the Jews
shared a physiognomic and spiritual uniformity, such that "one could
almost designate them a unique race." [441] Wellisch agreed with the popular
theory that modern Jews were made up of Ashkenazi and Sephardi racial
types. The Sephardim, who were believed to constitute about one-tenth of
all Jews, were descendants of the Semites, a race of "predominantly Oriental origins." The majority of Jews, the Ashkenazim, were thought to be more
closely related to the Aryans, but with a strong "Near Eastern" influence.[442]
Using the statistics from existing studies of over 5,000 Jews from across
Eurasia, Wellisch calculated the larger picture of each type's distribution:

---

[440] Ibid., 21.

[441] Siegmund Wellisch studied at the Vienna Technische Hoschschule. In 1889 he was an assistant in astronomy
and geodesy and, in 1893, began working at the Vienna Municipal Building Authority. In 1922 he worked for
a division of the Vienna Magistrate. See Geisenhainer, *Rasse ist Schicksal*, 198.

[442] Or, as Otto Reche claimed, the "Taurian race."

|          |          |          |          |            |
|----------|----------|----------|----------|------------|
| Ashkenazi | AB 7.1   | A 40.8   | B 18.7   | O 33.4     |
| Sephardi  | AB 5.0   | A 33.0   | B 23.2   | O 38.8 [443] |

By using a complex and unclear mathematic equation of his own design, which involved blood types and "gene ratios," Wellisch claimed that he was able to determine with which races the Jews had mixed and to what degree this had occurred, as well as what their original, "untouched" racial types had been. He believed that he had discovered the exact racial composition of the Jews and maintained that all were a mixture of the following: Near Easterners, Semitic Orientals, and Aryans (which included Nordic Amorites, Hunnish [Southwest Asian] Mongols, and Egyptian Negroes).[444] The proportion of each blood type within a Jewish population determined whether it was Ashkenazi or Sephardi. The results were startlingly precise: The Ashkenazim were described as a racial mix of "62 percent Near Eastern, 24 percent Oriental, and 14 percent Mongolian" blood. The Sephardim were made up of these same three types, but with differing proportions—"12 percent Near Eastern, 80 percent Oriental, and 8 percent Mongolian" blood.[445]

Wellisch also claimed to have identified the original racial types of the Ashkenazim and Sephardim and how they came to be different through "blood mixing." The Ashkenazim of Central and Eastern Europe were the descendants of Jews who had originally migrated northward from upper Palestine. Over the course of their travels, they had received a significant quantity of "Mongol blood" (*mongolisches Bluteinschlag*). By contrast, the Sephardim had migrated westward out of southern Palestine and had absorbed more of what Wellisch termed "Mediterranean" or "Negro blood" from the peoples of North Africa. The Jews' frequent "Aryan appearance" was the result of generations of miscegenation; "true Semitic blood," Wellisch claimed, "pulses in less than half of the mixed Jewish race" (*Echt semitisches Blut pulsiert in weniger als der Hälfte des jüdischen Rassengemenges*). Blood mixing accounted for the physiognomic diversity among modern Jews. Because they had not mixed with the "colored racial elements" of the Mongols

---

[443] Wellisch, "Serologische Untersuchungen über das Rassentum der Juden," 205.

[444] Ibid., 207.

[445] Ibid., 206. These figures are listed in the same, "Table 7," page 206, along with the "difference, margin of error, and blood-type formula." See also Table 8, page 206 "Blood Groups of Jews," and Table 9, page 207, "Near Easterners" and "Orientals."

and Negroes, the majority of Ashkenazim (84 percent, to be exact) were light-skinned.[446] Wellisch's research, impressive as it may have seemed with its facts and figures, basically introduced nothing new to existing theories on the race of the Jews. Italian physician Cesar Lombroso had similarly claimed that modern Jews were physically "more Aryan than Semitic," and race theorist W.Z. Ripley theorized that approximately nine-tenths of modern Jews were different from "original Semites" in terms of their cranial shape.

In the end, seroanthropology seemed no more accurate than physiognomy for anthropological classification of the Jewish people. Like conventional race theorists, Wellisch still concluded that Jews were predominantly Ashkenazim—a common claim, and viewed as problematic by many because of its generality. Wellisch, too, recognized the limitations of blood. "Currently," he admitted, "it is still difficult to give a definite serological response to the racial identity of the Jews." He blamed this mainly on the insufficient number of blood type analyses of Jews throughout Southern and Western Europe, which had "left much to be desired."[447] Statistics on the Sephardim were especially lacking—only 500 "definitely Sephardi" subjects had ever been examined.[448]

The fact that the German Institute for Blood Group Research published an article on the racial makeup of the Jews is not in and of itself an indicator of anti-Semitism, as anthropological studies of Jews were common. An anti-Semitic bias was, however, suggested by other articles in the institute's periodical. Wellisch's contribution was soon followed by a blatantly *völkisch* article on the unique "racial odor" of the Jews by race theorist Hans F.K. Günther. Günther had a reputation for blending racial prejudice with medical fact—this earned him the nickname *Rassengünther*, or "Race Günther," by his contemporaries. At the request of J.F. Lehmann, he wrote *Rassenkunde des Deutschen Volkes* (Racial Anthropology of the German Nation), which was first published in 1922. It was reprinted many times without any substantial changes and became the standard work of National Socialist racial doctrine.[449] His later contribution to the

---

[446] Ibid.

[447] Ibid., 205.

[448] According to existing serological studies, Wellisch further explained that the two types of Jews could be distinguished by various types of "blood type ratios." See Tables I–IV, ibid.

[449] Ute Deichmann, *Biologists under Hitler* (London: Harvard University Press, 1996), note 40, 389.

*Zeitschrift für Rassenphysiologie* was representative of these same ideals. Günther did not discuss the role of blood science in racial differentiation, but instead the notion that each racial type *smelled* a certain way—what he referred to as their "racial odor" (*Völkerodor*).[450] The peoples from Central Europe were described as having an almost "pungent and rancid" smell—"especially from their armpits." Günther maintained that this had been established through medical research; the sweat glands in the armpits of the Central Europeans were found to be "much larger" than those of the Japanese, whose were quite small and "only visible with a microscope." He was most preoccupied, however, with the *Völkerodor* of the Jews and cited numerous examples from Roman to modern times describing the "smell of the Jews," an "odor Judaeus," and how "even their small children smelled."[451] Günther claimed that his analysis was not meant to denigrate Jews, but the racism is unmistakable. The decision to publish such an article suggests that the German Institute for Blood Group Research shared these principles.

At the same time, the institute approved the publication of other articles that seemed to balance, or at least temper, this anti-Semitism by questioning the assertion that Jewish blood was different from that of non-Jews. A separate work, printed in the same edition of the *Zeitschrift* as Günther's, refuted Manoiloff's claim that the blood of different races could be distinguished through a special chemical reaction. At this point, Manoiloff was still holding fast to the claim that he could identify an individual's race from a blood sample alone "91.7 percent of the time." However, S.P. Grigorjewa, a fellow Russian physician, voiced strong reservations about Manoiloff's technique, because the chemistry of the reaction could not be explained. Grigorjewa tested it himself on approximately 1,000 subjects, nearly half of whom were Slavic, and the remainder largely Jewish or individuals "of mixed blood"—which usually referred to Russian and

---

[450] Hans F.K. Günther, "Der rasseeigene Geruch der Hautausdünstung," *Zeitschrift für Rassenphysiologie* 2 (1929/1930): 94–99.

[451] Günther's account describes how in ancient Rome, Ammianus Marcellinus reported to Emperor Marcus Aurelius (161–180 AD) on the "smell of the Jews" as he passed through Palestine to Egypt. According to Günther, the Grimm Dictionary described the term "Jüdern" as meaning "like a Jew or having a Jewish smell." Several early medieval Christian poets made similar references to an "odor Judaeus." Later, in the early eighteenth century, Schudt mentioned how "the Jews of Frankfurt and other places smell"—a trait believed to be inherited, as "even their small children smelled." Ibid., 98.

Ukrainian lineage.[452] To better examine the accuracy of Manoiloff's process, Grigorjewa retested each subject numerous times over the course of two months. He scrutinized thirty-two randomly chosen individuals even more closely: for six days, blood was drawn from the same subject at the same time every day. This careful repetition revealed that the resulting colors could vary from blue-green or violet to various nuances thereof that were significantly different from that individual's initial color reaction.

Grigorjewa's research certainly indicated that Manoiloff's chemical technique was not nearly as reliable as he had claimed. The blood of the Russian subjects yielded a characteristically Russian violet color in only 62.8 percent of the cases. Of the remaining cases, about a quarter of the blood samples had turned a shade of blue or green which, according to Manoiloff, indicated the presence of Jewish blood. None of the Russians in Grigorjewa's study, though, were Jewish. Furthermore, Grigorjewa found that 12 percent of the reactions were so imprecise as to prevent classification altogether. The results were similar among the Ukrainians, whose blood gave the expected reaction about 60 percent of the time, but had erroneously turned the "Jewish color" about a third of the time and had been indefinite for the remainder. The blood samples drawn from Jews turned the correct color less than half of the time (40 percent).[453] Seventy percent of the time, the color of mixed descent subjects' blood indicated that they were "racially pure" Russians; the reactions of those remaining indicated, again, that they were Jewish. Grigorjewa observed the most accurate results in his analyses of Russians, among whom the racial categorization and chemical response matched just over 60 percent of the time. The most unreliable results were observed among the Jews, whose blood often prompted an "irregular reaction."[454] The inconsistencies in his research, such as the Jewish reactions among non-Jews, or variable

---

[452] Among these were 443 "Slavs"—309 of whom were Russian, 128 Ukrainian, four Polish, one Serbian, and one Moldavian. The subjects examined also included 168 "Hebrews," seventeen Armenians, seven Greeks, one "Tatar," one German, two Lithuanians, and 261 individuals of "mixed descent." Grigorjewa tested Manoiloff's reaction on 900 subjects of various racial types. Nearly half were categorized as "Slavs"; also included were 168 Jews, twenty-eight individuals from throughout Central and Eastern Europe, and 261 subjects of "mixed descent," which, the author explained, usually meant that the individual was a mix of Russian and Ukrainian, or of Russian and/or Ukrainian with Jewish blood. See S.P. Grigorjewa, "Die Manoilowsche Reaktion als Mittel zur Rassenbestimmung beim Menschen." *Zeitschrift für Rassenphysiologie* 2 (1929/1930): 92–93.
[453] Ibid., 92.
[454] Ibid., 93.

reaction colors recorded over time with one individual, led Grigorjewa to conclude that Manoiloff's so-called "racial hormones," if they existed at all, were not constant.[455]

Other inclusions in the *Zeitschrift* suggested that the racial biases of its editors did not necessarily affect their judgment. Numerous acknowledgments, however small, suggested that they were more even-handed than would be expected. In the *Zeitschrift*, authors commonly recognized Jewish physicians for their contributions to blood science.[456] In an extensive volume on the clinical, legal, and racial applications of blood science, published in 1932, editor Paul Steffan readily acknowledged that the "most critical" contributions in serology had been made by Karl Landsteiner, Ludwik Hirszfeld, and Felix Bernstein, all of whom were of Jewish lineage. [457] Steffan praised Hirszfeld's work, which had prompted some of the most important anthropological questions since "the beginning of blood group research."[458] Oswald Streng, another member of the German Institute for Blood Group Research, referred to Hirszfeld's 1918 study as fundamental in its demonstration that blood type distributions varied across groups of people—a fact subsequently confirmed by "hundreds of researchers around the world."[459] Similar references were made throughout many of the institute's publications—what amounted to admitting the "Jewish roots" of their science. On one occasion, Reche even invited German Jewish blood scientist Felix Bernstein to participate in an institutional colloquium.

Though there were many prominent Jewish blood scientists at this time, such as Bernstein, Hirszfeld, and Schiff, none of them were members of the institute, even though the German Institute for Blood Group Research was the only one of its kind in Germany. It would stand to reason that they were excluded because of anti-Semitism. However, in 1928 Reche invited Karl Landsteiner, then employed at the Rockefeller Institute in New York, to become an "honorary foreign member." In the summer of that year, Reche

---

[455] Grigorjewa noted that the presence of malaria in the subjects, who were hospital inmates, may have distorted the results.

[456] Even Reche acknowledged Landsteiner and the Hirszfelds for their discoveries. See Otto Reche, "Blutgruppenforschung und Anthropologie," *Volk und Rasse* 3, no. 1 (1928): 2.

[457] See Steffan, ed., *Handbuch*, 385.

[458] Ibid.

[459] Oswald Streng, "Die Bluteigenschaften (Blutgruppen) der Völker, besonderers die der Germanen" *Festschrift für Hermann Hirt* 1 (1936): 412.

Figure 12.Karl Landsteiner. From Paul Speiser and Ferdinand G. Smekal,
Karl Landsteiner: *The Discoverer of the Blood Groups and a Pioneer in the Field
of Immunology; Biography of a Nobel Prize Winner of the Vienna Medical School*
(Vienna: Hollinek, 1975).

actually met with Landsteiner. (See Figure 12.) Paul Steffan later described
Reche's account of their meeting:

> I have an interesting piece of news! A few days ago I visited Landsteiner in
> New York! He is a tall, slim, attractive man with a slight scar on his left cheek;
> his racial type is not very obvious. He gave the impression of an individual
> completely and exclusively devoted to one topic—any divergences from this
> were somehow objectively related back to it. Landsteiner had just returned
> from a visit to Berlin, where he had met with Schiff. I mentioned to him that
> I found his relationship with Schiff to be mysterious. I don't think he appreci-
> ated my opinion. Landsteiner's work has been extraordinarily interesting. He
> has generated some very good ideas.[460]

When Landsteiner received Reche's offer, he initially declined, explaining:

---

[460] Geisenhainer, *Rasse ist Schicksal*, 135.

I hope you will understand that this decision was not an easy one for me and only came after a great deal of personal reflection. I have heard that Dr. [Fritz] Schiff, one of the most respected authorities in the area of blood group research, was not invited to join [the institute]. I believe that it might be awkward if I were to proceed independently of my esteemed colleague Schiff.[461]

The institute did not extend an offer of membership to Schiff, Reche explained, for two reasons: Schiff had "expressed differences" with their work, and he had attempted to "turn others against" the institute. Reche was also convinced that Schiff had similarly tried to influence Ludwik Hirszfeld, but he did not explain his reason for this suspicion.[462] Reche's selection of members was likely influenced by each individual's estimation of seroanthropology. Based on his own research in Berlin, Schiff had flatly rejected blood as a reliable means of racial differentiation. Landsteiner, however, had not discounted studies of blood for anthropological purposes.

After immigrating to the United States in the aftermath of World War I, Landsteiner proved himself a welcome addition to the Rockefeller Institute. Here his work was mainly with the therapeutic uses of blood, such as transfusions. By this time, the significance of his 1901 discovery of the blood types was fully realized, and on December 11, 1930, Landsteiner was awarded a Nobel Prize for his contribution.[463] In acknowledging the award, Landsteiner reflected briefly on the diverse uses of the blood types since their finding, which had mainly been in the areas of transfusion and law, as evidence in paternity and criminal disputes. When asked in 1930 about what had inspired him to detect differences between the blood of one individual and another, he explained that he had been motivated by the fact that

Differences exist in the blood of different animal species. I set out to examine the question whether individual differences are not present within a species. This led to the demonstration of individual differences, the blood groups in humans ... but this will not interest the layman.[464]

---

[461] Ibid., 134.

[462] Ibid., 134–135.

[463] "Landsteiner did not consider the 'sheer accident of discovery' of the blood groups and their elaboration as his greatest contribution to medical science. Although he derived great personal satisfaction from receiving a Nobel Prize, he would have preferred to be honored for his critical demonstration of the specificity of serological reactions, the title of his book." Gottlieb, "Karl Landsteiner, the Melancholy Genius: His Time and His Colleagues, 1868–1943," 18.

[464] Ibid., 26.

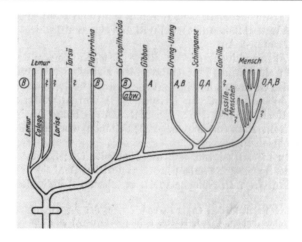

Figure 13. The blood types O, A, and B in "humans and several other primates," according to Landsteiner and Miller. Oluf Thomsen, "Serologie der Blutgruppen," in Paul
Steffan, ed., *Handbuch der Blutgruppenkunde* (Munich: J. F. Lehmann, 1932), 96.

Race was not a guiding factor in Landsteiner's initial research. He mentioned that the basic mode of the blood types' inheritance, a requisite
for racial characteristics, was only made years later by von Dungern and
Hirszfeld.[465] He further mentioned Hirszfeld's "remarkable observation"
that there were characteristic blood type patterns according to race.

By the time of his award, Landsteiner judged the volume of seroanthropological studies to be "nearly impossible of complete review."[466] Although
his primary interest was in what he referred to as "practical medicine" and
blood, Landsteiner had carried out seroanthropological research. In 1925
he co-authored an article with Charles Miller at the Rockefeller Institute
that compared the blood types of various primates to those in humans.[467]
According to their blood tests of white and "Negro" subjects, it seemed that
certain blood types were affiliated with specific primates—such as chimpanzees with types A and O, or orangutans with A and B, though there
were no differences pronounced enough to permit any direct correlations.
It was only clear that the blood of "whites and Negroes" was not as different
as that of "human blood and that of anthropoid apes."[468] (See Figure 13.)

---

[465] Karl Landsteiner, "Individual Differences in Human Blood," *Science* 73, no. 1894 (1931): 406.
[466] Ibid.
[467] Ibid., 407.
[468] Karl Landsteiner and Charles Philip Miller, Jr., "Serological Studies on the Blood of the Primates: The Blood
Groups in Anthropoid Apes," *Journal of Experimental Medicine* 42 (1925): 853–862.

In the following year, Landsteiner co-discovered what was referred to as the MNP system in the blood. Like the blood types, the MNP factors were based on the presence or absence of certain proteins on the surface of the blood cells. These, too, were genetic and, as Landsteiner concluded, could be used to broadly differentiate between races, as their incidence fluctuated noticeably among "Caucasians, blacks, and Native Americans."[469]

Clearly then, despite his professed primary interest in "practical" applications of blood science, Landsteiner carried out seroanthropological research and, even more importantly, he was not critical of its basic premises, as his acquaintance Schiff had been. On the contrary, Landsteiner's work delved into areas of seroanthropology of which his European colleagues were still unaware. His contributions had the added advantage of coming from the discoverer of the blood types themselves, who was now a Nobel Prize winner to boot. It was for these reasons that Reche invited Landsteiner to join the German Institute for Blood Group Research.

Obviously, Reche did not want his members to openly discredit or too strongly criticize seroanthropology. Provided they refrained from this, he extended membership to those of Jewish descent when it served his purposes. This is not to say that Reche was not an anti-Semite, because he most certainly was, but he was also a pragmatic opportunist, and some of the best blood scientists of the day were Jewish. Even though Reche strongly suspected that Siegmund Wellisch was Jewish, Wellisch was a frequent contributor to the *Zeitschrift für Rassenphysiologie*. In the summer of 1929, Reche wrote to Steffan: "Wellisch is a brilliant (*grossartige*) contributor... if we only knew his racial type (*Menschengruppe*)!"[470] Reche had questioned his colleagues in Vienna to this effect but had not yet received a response. Despite these doubts, Wellisch remained a member of the institute, certainly because his analyses usually made a positive correlation between blood and race.

In the first edition of the *Zeitschrift für Rassenphysiologie*, Wellisch authored an article on the blood relations of people and races in which he claimed that the blood types were, "without a doubt," a suitable way to distinguish between racial types.[471] Furthermore, he reinforced his claims

---

[469] Steffan, ed., *Handbuch*, 15.
[470] Quoted in ibid., 198.
[471] Siegmund Wellisch, "'Blutsverwandschaft' der Völker und Rassen," *Zeitschrift für Rassenphysiologie* 1 (1928/1929): 21.

with theories and formulas that appeared scientifically grounded. Wellisch proposed a new theory concerning the separate "zones of origin" of blood types A and B in which the distributions of type could be mathematically analyzed to indicate whether a subject group was closer in origin to the A- or B-type race.[472] Provided no miscegenation occurred, he declared that this equation would reveal a quantitative relationship that would remain unchanged throughout successive generations.[473] Wellisch further explained how his theory of "agglutination poles" corresponded with German race theorist Eugen Fischer's contention that there had originally been three different distinct racial types in Europe—the Germans, Celts, and Slavs. According to Fischer, these peoples had migrated in varying degrees away from the native land of the Indo-Germans. By comparing the incidence of type A to B blood, Wellisch's formula indicated which of the three types a group was most closely related to.[474]

Wellisch's interpretations not only defended the merits of seroanthropology but also had the added advantage of confirming that the German people were "racially Nordic". Based on their blood types, Wellisch placed Germans closest to the (blood type) "A pole." By contrast, with their higher levels of type B, the Slavs were farthest from it, and the Celts somewhere in between.[475] Included among the other groups closest to the A pole with the Germans were those traditionally categorized as Aryan—the Angles, Swedes, Norwegians, English, and Dutch—whereas the Slavic types included Poles, Lithuanians, and Russians.[476]

Even though he acknowledged that there were still "gaps in the research," Wellisch maintained that the Angles of the East Schleswig area of Schwansen and the indigenous peasant Germans of Prüm were "just as Aryan as the Swedes."[477] He claimed that blood indicated simple, definitive results. Just as the Magyars could not deny their Mongolian origins,

---

[472] Ibid., 28.

[473] Ibid., 24.

[474] Ibid., 30.

[475] Included amongst those closest to the A pole were also "Austrians, Serbs, Macedonians, Gypsies (Roma Sinti), and Indians." Those groups classified as "Slavic" were eastern Germans, Finns, Hungarians, Poles, Lithuanians, Russians, Koreans, Chinese, Ainu-Japanese, and Manchurians. The "intermediate" Celts included Belgians, French, Italians, North Africans, Sephardim, Egyptians, Senegalese, and groups from Madagascar and the Congo.

[476] Ibid.

[477] Ibid., 31.

he explained, "neither could the Egyptians deny their Negro and Oriental descent, nor the Gypsies their close relations with Indians."[478] In a separate analysis, once again in the *Zeitschrift*, Wellisch placed the Nordic Aryans and Indians, represented respectively by types A and B blood, at opposite ends of a racial spectrum.[479] Based on the invitation extended to Landsteiner, the repeated praise given physicians of Jewish descent, and the inclusion of Wellisch's work, in spite of strong reservations about his racial makeup, it would seem that an individual's defense of seroanthropology as a valid science was more of concern to the officers of the institute than his "racial type."

While there were hints of anti-Semitism, Reche's work published before 1933 does not suggest that pinpointing the racial identity of the Jews was a main priority. His references to Jews alternately as *Homo mediterraneus*, *Orientalis*, and *Semitisches* revealed his conviction that they were a "non-German" race, but he did not aggressively pursue the matter.[480] As for the institute's exclusion of Fritz Schiff, one must take into account the fact that Schiff had voiced serious doubts about the usefulness of seroanthropology. In 1929, after directing yet another study on the blood types of Berliners, Schiff did not change his perspective.[481] He had taken the necessary precaution of gathering a large, adequately representative set of subjects but again emphasized the lack of correlation between blood and race. Schiff also criticized the notion that there was some relationship between blood type and mentality, and referred specifically to Boehmer and Gundel's affiliation between type B blood and criminal behavior. He rightfully pointed out that type B blood was also more common in eastern Germany than elsewhere in the country. "Are we to conclude then," he mocked, "that Prussians are more criminally inclined than Rhineland Germans because they have a higher incidence of type B blood?"[482] In spite of his earlier protests, Landsteiner did join the German Institute for Blood Group Research

---

[478] Ibid., 33.

[479] Siegmund Wellisch, "Die Analyse der Dreirassentheorie," *Zeitschrift für Rassenphysiologie* 1 (1928/1929): 66–71.

[480] Geisenhainer, *Rasse ist Schicksal*, 188.

[481] Fritz Schiff, "Zur Serologie der Berliner Bevölkerung," *Klinische Wochenschrift* 8, no. 10 (1929): 448–450. Schiff's 1924 study examined 750 patients; there were 2,500 subjects in his 1929 research. Schiff did acknowledge that further serological research could disprove any of his conclusions. Ibid., 448.

[482] V. Myslivec, *A Contribution to the Study of Human Blood Groups* (London: Royal Anthropological Institute of Great Britain and Ireland, 1941): 70.

in 1928, at which point Reche announced in the *Zeitschrift* his pleasure that the "discoverer of the blood types" was now a member.[483]

## VÖLKISCH PROPAGANDA

Despite the lack of biological proof, anti-Semitic propaganda persisted in referring to differences between Jewish and German blood. In 1927, in a call for donors, *Der Stürmer* described the inferiority of Jewish blood:

> The unemployed in Fürth have just as much blood as those in Nuremberg and other cities. On the other hand, the Jews of Fürth have blood to donate as well. Due to their inferiority, however, their blood is not needed. Jewish blood is a combination of Negro, Mongolian, and a portion of animal (probably ape) blood. A transfusion of such blood into the body of a non-Jew would have a poisonous effect.[484]

Of course, there was no medical evidence to substantiate such claims. Sim-ilarly, there were no findings to suggest that transfusions between different racial types were harmful. It was not true that that transfusing "non-German" blood into a German would poison the recipient. Lethal "poisoning" could, in fact, take place in transfusions, but this was only a result of blood type incompatibility, and blood type, not racial type, was used to differentiate between donations during the Weimar Republic.

In theory, race propagandists could have exploited seroanthropology. It was exceedingly rare for anti-Semites to draw upon blood type studies in their efforts to reinforce ideas of racial difference—though blood science was, on occasion, discussed in more "intellectual" anti-Semitic writings. The popular *Handbook on the Jewish Question* (Handbuch der Judenfrage), which was reissued forty-nine times between 1907 and 1944, first commented on seroanthropology in 1931.[485] In a chapter devoted to the "racial anthropology of the Jews," reference was made to the question of serology and the close blood relations observed between the "Eastern, Middle Eastern, Southern Jews, and Oriental races." Wellisch's observations concerning

---

[483] Otto Reche, "Mitteilungen des Vorstandes der Deutschen Gesellschaft für Blutgruppenforschung," *Zeitschrift für Rassenphysiologie* 1 (1928/1929): 5.

[484] *Der Stürmer* (September 30, 1927).

[485] This handbook was originally published in 1887. See Theodor Fritsch, *Antisemiten-Katechismus* (Leipzig: Hermann Beyer Verlag, 1887).

Ashkenazi and Sephardi Jews were also mentioned, specifically his claim that each type had both "Middle Eastern and Oriental blood" and that the proportions of types A and B blood could be used to determine whether a group of Jews was primarily of Ashkenazi or Sephardi origins. Propagandists likely did not refer to clinical studies of blood and race because of the simple fact that there were only four blood types, and these existed in all peoples. This did not, however, prevent occasional distortions or even outright fabrications of medical fact with reference to blood science.

In his 1931 *The Myth of the Twentieth Century* (*Der Mythus des 20. Jahrhunderts*), Arthur Rosenberg made frequent use of blood rhetoric in his racial descriptions. His pseudo-medical language was like that in *Der Stürmer*:

> A new faith is awakening today: the faith that blood will defend the divine essence of man; the faith, *supported by pure science* [italics added], that Nordic blood embodies the new mystery which will supplant the outworn sacrament.[486]

Of course, the supporting "pure science" is not cited, but it was presented as fact that physiological blood defilement could occur from sexual contact with a "racial other," or by receiving the racial other's blood via a transfusion.

In 1932, a small notice detailing two scenarios of blood defilement appeared in *Der Hammer*, the monthly periodical of the German Nationalist Workers' League.[487] The first example reiterated the storyline of *The Sin against the Blood*, summarizing how the blood of an "Aryan woman" had been permanently defiled by a single sexual encounter with a Jew. As a result, she later gave birth to a child who appeared "completely Jewish"—even though his biological father had been German. The second example described how the blood of a "racially pure" Jew was transfused into a patient of German origin. The passage reported on a 1919 incident in a clinic in Mainz in which six anemic women received blood transfusions from healthy men. Surprisingly quick improvement occurred in five of the cases.[488] In the sixth case,

---

[486] Arthur Rosenberg, *The Myth of the Twentieth Century: An Evaluation of the Spiritual-Intellectual Confrontations of Our Age* (Torrance, CA: Noontide Press, 1982) (first English edition).
[487] *Der Hammer* also appeared as a supplement to the *Ostdeutsche Rundschau*.
[488] G. Lindeman, "Über Blutüberpflanzung in der Geburtshilfe und Gynäkologie," *Münchener Medizinische Wochenschrift*, no. 11 (1919): 285–286.

though the patient received only 50 ccm of blood (versus the 100 to 400 ccm of the others), she had a "very negative" reaction. After the transfusion, the woman began to shiver and her pulse dropped. Perhaps it was of significance, the author mused, that in this instance the German woman had received the blood of a … "racially pure Jew!!!" (*reinrassigen Jüdin*). "Unfortunately," the article continued, this had—unbeknownst to the patient— "infected [*infiziert*] her with inferior blood." Like the German woman who had had sexual relations with a Jew, the transfusion would effectively contaminate the blood for generations, preventing "true German" offspring.[489]

The transfusion in Mainz was also mentioned in a *Der Stürmer* article entitled "Jewish Blood—A Startling Scientific Finding."[490] This account further detailed how the donor had been quite anxious about the results, because they proved that Jewish blood was "poisonous" for Germans. The article warned:

The Jews know that if the Germans were to find this out, they would then recognize the enormous racial difference between Germans and Jews. Perhaps only then would German women and girls avoid the Jewish pest and not receive his poison into their bodies.[491]

Dr. Hentschel, the author, voiced his agreement. He confirmed that blood typing was necessary for transfusion compatibility—an accurate statement—but then contradicted this practicality by advising that blood exchange between individuals "completely dissimilar in appearance" not be permitted. In fact, both accounts were based on an actual article in the medical periodical *Münchener Medizinische Wochenschrift*, which did describe an anemic woman's unexpected adverse reaction to a relatively small transfusion of blood. Her response, which *Der Hammer* and *Der Stürmer* conveniently neglected to mention, was simply a result of blood type incompatibility. The original report did not even mention race.

---

[489] Willibald Hentschel, "Rassenverschlechterung durch Juden," *Der Hammer*, no. 725/726 (September 1932): 246.

[490] "Judenblut: Eine aufsehenerregende wissenschaftliche Festellung," *Der Stürmer* 10, no. 31 (December 1932): 3.

[491] Ibid. No first name or initial are given for Hentschel. This was certainly Willibald Hentschel (1858–1947), who has been referred to by one author as "one of the most eccentric but influential propagators of racial hygiene and rural romanticism in the Wilhelmian and Weimar eras." Hentschel co-founded the periodical *Der Hammer* in 1903. Richard S. Levy, *Antisemitism: A Historical Encyclopedia of Prejudice and Persecution* (Oxford: ABC-CLIO, 2005), 296.

JEWS AND SEROANTHROPOLOGY

German Jews not affiliated with Zionism vehemently objected to the growing scientific fiction of "Jewish blood" and strongly criticized all relevant
research. In 1925 German Jewish psychologist Kurt Lewin wrote in the
*CV-Zeitung*, the weekly publication of the assimilationist group the Central
Committee for German Citizens of the Jewish Faith, that "anti-Semitic race
theorists have still not recognized that blood type cannot be definitively
linked to race." It was not possible, he reiterated, to differentiate "a Jew from
a German" by blood type. Lewin even expressed reservations concerning
the heredity of the blood groups themselves; what if these, he posited, like
physiognomic characteristics, were not also affected by "climate, lifestyle,
or other environmental factors?"[492] Several years later, he noted that ongoing efforts to align blood and race had proved "completely worthless." Like
Lewin, liberal German politician Paul Nathan also criticized misappropriations of seroanthropology. He declared that it was not his intent to let his
own political beliefs affect his interpretation of medical findings or prevent their distribution—regardless of their implications. To Nathan, however, a self-proclaimed "layperson in this area," it seemed that seroanthropology worked "to divide humanity and elevate certain racial types among
others."[493] These men were not alone in recognizing that select studies of
blood and race were biased. For this reason, studies that acknowledged the
science's shortcomings were well-received. Schiff's 1924 study was favor-
ably referred to by physicians Franz Schütz and Wolfgang Kruse, both
respected blood scientists. Kruse wrote that the Jews "approximated their
host population in their blood group distributions."[494] Ironically, though,
physicians of Jewish descent had played a central role in the development
of racial studies of blood. Were they somehow to blame for the distortion
of blood science by political extremists?

There had been no significant discussions of using blood to analyze race
prior to Ludwik Hirszfeld's landmark 1918 study. Hirszfeld set the stage
for a movement to prove that racial types could be differentiated through

---

[492] *Centralverein Zeitung* (1925): 43.

[493] To indicate at least a cursory knowledge of developments in the field, Nathan did mention various medically
based blood type studies in a footnote. Among the works referenced was that of Schiff and Ziegler.

[494] W. Kruse, "Über Blutzusammensetzung und Rasse," *Archiv für Rassenbiologie* 19, no. 1 (1927): 20–33.

blood. Anthropological studies of blood, however, were not his sole focus, and after the war Hirszfeld returned to his native Poland to resume general studies of blood. In Warsaw Hirszfeld established an institute of serology. In relatively quick succession, he was appointed deputy director and scientific head of the State Hygiene Institute and, in 1924, professor. In 1931 Hirszfeld was named full professor at the University of Warsaw, in addition to variously heading the Epidemiology Institute, the Department of Bacteriology and Experimental Medicine, and the Serum Section of the State Institute of Hygiene and the Institute of Science. Under Hirszfeld's guidance, Warsaw became an international center for blood science and seroanthropological study.[495]

In spite of his many achievements, Hirszfeld was most renowned for his work during World War I. Philip Levine, an American Jewish serologist, still remarked that "there was no question that the Hirszfelds' work began the application of anthropology towards blood science."[496] Even years before World War I, Hirszfeld had entertained the idea that there might be some connection between blood and race. Given the opportunity to study this relationship, he had wasted no time in coordinating an analysis of the blood types of the various ethnic types accessible to him. From a modern perspective, it would certainly seem ironic that a racially unbiased physician, and particularly one who would have been sensitive to the realities of racial anti-Semitism, would be so eager to introduce a more effective means of racial identification.

The very premise of Hirszfeld's wartime blood type study indicates his belief in qualitative racial differences. Hirszfeld's Salonika study has since been criticized as racist, and this does seem to be suggested in his interpretation of his findings, in particular the portrayal of types A and B as racial opposites.[497] In addition, Hirszfeld generalized the traits of his different

---

[495] Frank Heynick, *Jews and Medicine*, 438.

[496] Philip Levine, "Menschliche Blutgruppen und individuelle Blutdifferenzen," *Ergebnisse der Inneren Medizin und Kinderheilkunde* 34 (1928): 146.

[497] Myriam Spörri suggests that the Hirszfelds' analysis and presentation of their results were racially biased. As she points out, their 1918 diagram of the various serological distributions refers only to types A and B out of the four different types, a clear indication of their polarity. Differences between the two are further emphasized by their representations in a line graph: type A is only slightly shaded, and B is solid black. Were these colors coincidental? The study predominantly associated type A blood with the white, European peoples, whereas the dark-colored races were linked to B. Spörri pointed out that even the label of "A," as the first letter of the alphabet, could be construed as yet another marker of the (racial) superiority of the Western Europeans. The biochemical index, presumably the most innovative component of the study, also seems to place more "value" in type A blood. Those racial types with more type B blood, traditionally viewed by Nordic

subject groups, explaining that they modified their approach according to ethnicity in order to collect the necessary samples:

> We had to speak in a different way to each nation. It was sufficient to tell the English that our objectives were scientific. We allowed ourselves to fool our French friends by telling them that we could determine with whom they could sin with impunity. We told the Negroes that the blood tests would show who deserved leave; immediately, they willingly stretched their black hands out to us.[498]

The English are described as neutral and rational, while the French needed only the justification of sexuality to cooperate. The cowardice of the Russians is implied by the observation that many fainted, even though examiners "punctured only one finger and collected only a few drops of blood."[499] The Hirszfelds categorized their subjects into different national types; however, authors Lisa Gannett and James Griesemer have called attention to the fact that these classifications were heterogeneous. Some categories, such as the Jews and Arabs, touched upon ethnicity. In other cases, religion played a key role—as in the case of the Turks, who were labeled "Mohammedans." These are mainly national stereotypes, and while they are not necessarily any less excusable than racial bigotry, an important distinction nevertheless remains. Hirszfeld's apparent belief that different racial types existed is not reason enough to assume his belief in a racial hierarchy in which certain peoples were ranked above or below others. Hirszfeld's attempt to contribute positively to the discipline of racial anthropology, a completely legitimate science at the time, should not be interpreted as indication of racial discrimination on his part.

What may now seem like biased research was largely a product of the intellectual norms of this era, Hirszfeld's schooling in race theory, and, to a certain degree, his subsequent employment in Germany. After receiving his medical doctorate from the University of Berlin in 1907, Hirszfeld assisted in the serological department of the Institute for Cancer Research in Hei-

---

ideologues as anthropologically inferior, were also fittingly given lower biochemical indices; the Hirszfelds theorized that groups with indices under 1 were of African or Asian descent. "Mediterranean" types were those peoples between 1 and 2.5, while the label "Western European" was reserved for populations with a higher index—between 2.5 and 4.5. See Spörri, "'Reines Blut,' 'gemischtes Blut,'" 220.

[498] Ibid., 218.

[499] Leo J. McCarthy and Mathias Okroi, "The Original Blood Group Pioneers...the Hirszfelds," *Blood Banking and Transfusion Medicine* 2, no. 1 (2004): 25.

delberg.[500] His experience was not entirely uncommon, as non-Germans interested in a medical career were attracted to Germany's prestigious medical establishment. Racial anthropology and eugenics were respected areas of study in Western medicine and were even more strongly emphasized in Germany. As one author has described it, "medicine was suffused with the language of race science, and articles in medical journals discussed particular diseases or pathological states as though they were racially determined."[501] This influence would affect Hirszfeld even after he left Germany. He would also have been aware of a preference for physiognomic traits, as well as increasing questions about their reliability. All of these factors shaped Hirszfeld's decision to apply blood science to racial anthropology.[502]

References that Hirszfeld made to such ambiguities as "pure races" and "blood mixing" were common in anthropological studies of the time. His further division of humanity based on the geographic location of the blood types, and his creation of the "biochemical race index" to provide information on where a subject group was situated within a racial spectrum, do seem remarkably similar to *völkisch* studies of race. However, because of their studies in the science faculties of the universities and their heavy representation in the medical profession, Jewish scientists were well situated to employ the discourse and methodology of race science and ethnography.[503] When considered in this context, Hirszfeld's perspective was noth-

---

[500] Ibid., 25.

[501] Efron, *Defenders of the Race*, 11.

[502] Ludwik and Hanna Hirszfeld made a negligible attempt to record physiognomic traits of their subjects. When 500 blood samples were drawn from English subjects, the subjects' hair and eye color were also noted—as were the hair and skin color of the 500 French. The simple pigmented traits the Hirszfelds noted among the English and French did not suggest that blood group could be related to appearance. Furthermore, Ludwik Hirszfeld had already tested this association years earlier in his work with von Dungern, except on dogs, and no connection appeared to exist; he found that anatomically different dogs could share the same blood type (they were "biochemically similar") or, conversely, physically similar dogs could have different blood types. Young dogs with "the structure of the body and the coloring of their mother" would have the blood group of their father, and so on. To Hirszfeld, this suggested that the heredity of blood types and that of physical traits were entirely independent from each other. The same seemed to be the case in his subjects in Salonika. Again, none of the physical characteristics suggested any pattern specific to blood type. The Indians ("Hindus"), whom the Hirszfelds considered to be "anthropologically most similar to the Europeans," were, in fact, completely different in their blood type distributions; statistically, Indians had much higher frequencies of type B blood than any of the European groups. At the same time, Russians and Jews, who were traditionally considered to be so different from each other in "anatomical characteristics, mode of life, occupation, and temperament," had the exact same proportion of types A and B blood. There was no discernible pattern. The dogs, Hirszfeld pointed out, were not always "racially pure." See L. and H. Hirszfeld, "Des méthodes sérologiques au problème des races," *L'Anthropologie* 29 (1918–1919): 508.

[503] Efron, *Defenders of the Race*, 7.

Figure 14. Ludwik Hirszfeld. From Paul Speiser and Ferdinand G. Smekal,
*Karl Landsteiner: The Discoverer of the Blood Groups and a Pioneer in the Field
of Immunology; Biography of a Nobel Prize Winner of the Vienna Medical School*
(Vienna: Hollinek, 1975).

ing out of the ordinary. In this sense, he could be referred to as a "racialist"
rather than a racist.[504] (See Figure 14.)

Hirszfeld was not the only foreign physician to be influenced by education and employment in Germany. In fact, the origins and diffusion of
seroanthropology were largely traceable to German influence. Arthur
F. Coca, the editor of the *Journal of Immunology* in the United States and
the first researcher to note the high frequency of type O blood in Native
Americans, had also been mentored by Emil von Dungern in Berlin, with
Hirszfeld, in the early 1900s.[505] Some of the prominent early Asian seroanthropologists had also received their medical training in Germany.[506] These

---

[504] Racialism holds that inherited characteristics possessed by human beings permit us to divide them into a
small number of races so that members of these races share characteristics with each other that they do not
share with members of any other race. Kwame Anthony Appiah, "Racisms," in Larry May et al., eds., *Applied
Ethics: A Multicultural Approach* (Upper Saddle River, NJ: Prentice-Hall, 2001), 472–473.

[505] A.F. Coca and O. Deibert, "A Study of the Occurrence of the Blood Groups among American Indians," *Journal of Immunology* 8 (1923): 487.

[506] Such as Tanemoto Furuhata and Balcian Liang. See Schneider, "The History of Research on Blood Group Genetics: Initial Discovery and Diffusion," 286.

physicians then instructed others on the methodology and interpretation of seroanthropology. Nicholas Kossovitch, an influential blood scientist in interwar France, served on the Macedonian front during World War I, under the tutelage of Hirszfeld.[507] Like his mentor, Kossovitch was convinced that the blood types provided important insight into racial identity. As with Hirszfeld and many others who researched blood and race, Kossovitch's frustrations with physiognomic racial indicators led him to seek out a non-physical gauge of race. After conducting anthropological research in Northern Africa, Kossovitch referred to physiognomic characteristics as "useless" in differentiating between Arabs and Berbers.[508] Both of these groups, he explained, were of the "white race," with similar heights, cephalic indices, nose and face shapes; only after testing their blood was a difference revealed. As a further example of the "fixity of the blood groups," Kossovitch pointed to the Hovas, an isolated race of people in Madagascar.[509] Jan Mydlarski, a physician who worked with Hirszfeld in postwar Poland, also published extensively on his own independent studies of blood type and race.[510] In fact, it was his extensive study of approximately 11,000 Polish soldiers that first suggested a correlation between types A and B blood and "Western and Eastern" appearances, respectively.

The influence of German medicine in blood science was apparent even in such basic materials as testing supplies. Test serum was generally used in determining blood type. The Robert Koch Institute in Berlin, a provider of sera, distributed materials throughout Germany: the Greifswald University Pediatric Clinic, the Viral Research Department at the University of Jena, the National Medical Research Office, the Pathological Institute at

---

[507] N. Kossovitch, "Les groupes sanguins chez les Tschéques," *Comptes Rendus de Société Biologique* 93, no. 35 (1925): 1343–1344, and N. Kossovitch, "Récherches sur la race arménienne par l'isohémagglutination," *Comptes Rendus de Société Biologique* 97, no. 19 (1927): 69–71. "Kossovitch, a native Serb, went to Paris after World War I, where he obtained a position at the Pasteur Institute. He secured a position at the École d'anthropologie in 1931 and was the first in France to conduct research comparable to that of the Hirszfelds." William H. Schneider, *Quality and Quantity: The Quest for Biological Regeneration in Twentieth-Century France* (Cambridge: Cambridge University Press, 1990), 225.

[508] R. Dujarric de la Riviere and N. Kossovitch, "Les groupes sanguins en anthropologie," *Annales de Médecine Legale, Criminologie, Police Scientifique et Toxicologie*. (1934): 285.

[509] Ibid.

[510] J. Mydlarski, "Vorläufiger Bericht über die militäranthropologische Aufnahme Polens," *Kosmos* 50 (1925): 530–583. J. Mydlarski, "Probleme der Konstitution in Lichte der Anthropologie," *Polska Gazeta Lek*, no. 5 (1926). J. Mydlarski, "La différenciation du sang corrélative aux races humaines," *Institut Internationale D'Anthropologie*, 2. sess. Prague (1926): 189–199.

Wiesbaden City Hospital, and the Hygiene Institute of Albertus University were all clients. International requests were also forwarded from France, the Netherlands, Italy, and South America.[511]

Seroanthropological research in Germany and elsewhere had been substantial in the interwar period. By 1927, one physician reported that the blood types of 92,144 subjects, comprising "198 different racial types," had been examined.[512] By the end of the Weimar Republic, it was estimated that 1,000 seroanthropological articles, based on over half a million individual tests, had been published.[513] Much to the dismay of German race extremists, however, repeated efforts to discern the blood of one race from another—much less that of Jews from non-Jews—were not successful. As with other groups studied, the Jews' biochemical racial indices were not consistent and offered no new insight into the matter of their contentious racial classification. Furthermore, attempts at individual racial categorization through blood, as with Manoiloff's foray into assessing "racial hormones," had been both unreliable and inexplicable. As one French physician bluntly put it, "everyone knows we still do not have a method that allows the identification of an individual through their blood."[514] With little left to work with, references to clinical seroanthropology in racial propaganda were made, but they were relatively few.

Even with the impressive statistics that had been gathered over the course of the Weimar Republic and the seemingly tangled web of inconsistencies and unanswered questions they had created, on the eve of the Third Reich there still remained those who believed seroanthropology had potential. Siegmund Wellisch, for instance, stubbornly clung to the idea

---

[511] R86/2792.

[512] Of these, 56,922 were Europeans. Scheidt specifically listed which Europeans were studied (his nomenclature): Swedes 1,384, Norwegians 136, Scandinavians in general 138, Frisian Islands 800, Lapland 183, England, Scotland, and Ireland 1,015, Denmark 662, Holland 200, Germans 15,500, Swiss 543, Germans in Hungary and Romania 1,191, France 500, Italy 3,059, Poles 11,588, Slovaks 461, Russia 4726, Hungary 1,874, Romanians 5,325, Bulgarians 872, Serbs 500, Turks 527, Greeks 359, Middle Easterners 387, Egyptians 417, North Africa 500, Jews 1,875, Gypsies 385. Scheidt, Walter, "Rassenunterschiede des Blutes," 1927, Georg Thieme, Leipzig, Germany, 26.

In 1926, Hirszfeld also referred to numerous mass racial surveys conducted as of that date (Staquet, Belgians; Weil, French; Clairmont, German Swiss; Hoche and Moritsch, Viennese; Kossovitch, Chechens; Streng, Finns; Bruynoghe and Walravens, Congolese; Paseual, Filipino; Burton, Southern Australians; Nigg-Clare, Indians. See Walter Scheidt, *Rassenunterschiede des Blutes* (Leipzig: Georg Thieme, 1927), 25–26.

[513] Schneider, "Chance and Social Setting," 558.

[514] Paul Moureau, "Contribution à l'étude des facteurs d'individualisation du sang humain et leurs applications en medicine légale," *Revue Belge des Sciences Medicales* 7, no. 3 (March 1935): 178.

that more useful anthropological knowledge might be gathered from additional blood type research. As with all mass statistics, he noted, every additional study had the potential to alter, if not completely change, all previous findings.[515] Like Reche, he asked that the present findings in seroanthropology not discourage further interest—particularly in light of the inadequacies in both seroanthropology and racial anthropology in general. Even Grigorjewa, whose own research with Manoiloff's technique effectively exposed its uselessness, did not feel it should be altogether abandoned. He remarked on the "lingering possibility" that the diseased (malarial) subjects in the study may have affected the results, or that the nature of the so-called "racial hormones" was not yet completely understood. Still others theorized that the link between blood and race would become apparent only after even more extensive comparisons with other racial characteristics—a "prevailing belief" that Landsteiner agreed with.[516] In spite of such pleas and advice, the imprecision of seroanthropology would stymie its progress during the Third Reich, when the state mandated individual classification. The Nazis' persistent use of blood rhetoric in their propaganda, as well as its eventual incorporation into their legislation, made its abandonment that much more ironic.

---

[515] Wellisch, "Blutsverwandtschaft der Völker und Rassen," 21.
[516] Landsteiner, "Individual Differences in Human Blood," 406.

# CHAPTER VI

# BLOOD AS METAPHOR AND SCIENCE IN THE NUREMBERG RACE LAWS

Hitler's eugenic and racial beliefs attracted right-wing political and medical ideologues long before his appointment as chancellor in 1933. In 1930 German race theorist Fritz Lenz lauded him as the first politician "of truly great import, who has taken racial hygiene as a serious element of state policy."[517] The National Socialist Worker's Party was arguably the first political party in which racial hygiene was a central part of the political platform. When they did seize power, the Nazis immediately began to apply these ideas by separating "real" Germans from those categorized as "non-German." Because of the Nazi conviction that race was a biological category, the cooperation and support of physicians was deemed essential. To ensure this, *Gleichschaltung* forced all medical institutions, propaganda programs, and policies to serve the aims of Nazi selectionist racial ideology. [518] The Nazis' objective had consistently been to apply their principles under the pretense of legal, bureaucratic, orderly measures; with the guidance of doctors and lawyers, eventually all "racially other" individuals would be stripped of their civil rights. The segregation of Germany's predominantly assimilated Jewish minority was dictated by the Nuremberg race laws introduced in 1935. The legislation gave complicated guidelines, which replaced any previous state directives, for the classification of Jews.

The Nuremberg legislation depended on differences in "blood." Racial categorization hinged not on whether one was "Nordic" or "Aryan" but on

---

[517] Quoted in Proctor, *Racial Hygiene*, 61.
[518] Weindling, *Health, Race and German Politics*, 490.

whether one had "German blood." This terminology meant that the metaphorical notion of blood had to be quantified; how much "non-German blood" did an individual have to have in order to be considered Jewish? Or, conversely, how much "German blood" did a Jew need in order to "pass" as German? Imprecise categories of blood difference, whether portrayed in a sexual or scientific context, were used not only to construct racial categories, but also to interpret infractions of the legislation that resulted in a charge of blood defilement. Such references had been common in National Socialist discourse from the start. When the party issued its official twenty-five-point program in February 1920, Point Four stated that only those of "German blood" were considered *Volksgenossen* (German nationals), and only *Volksgenossen* could be German citizens. Hitler's *Mein Kampf*, written while he was briefly imprisoned during the early Weimar Republic, is replete with blood-based metaphors. Using this same blood rhetoric in legal categorization during the Third Reich would reveal the tensions of basing legislation in what were usually propagandistic ideals. A close analysis of the 1935 Nuremberg "Blood Protection Law" and its application will demonstrate the disconnect between medical and political notions of differences in blood. This disparity became increasingly pronounced during the prewar years of the "racial state" of the Third Reich.[519]

## SEROANTHROPOLOGY IN 1933

Even a cursory overview of seroanthropology in the early 1930s would have shown that it was ineffective as a tool of racial classification. Studies repeatedly demonstrated that no one blood type could be definitively linked to one racial type; patterns could only be broadly observed through differences in group blood type frequencies. The general public, however, was largely unaware of such technicalities, as seroanthropology, a new, very specialized science, had received little attention. This lack of awareness, combined with ongoing loose references to "blood differences" in the much more race-conscious environment of the day, led to frequent uncertainties and misconceptions. In a 1933 article entitled "What Blood Tells," American

---

[519] Michael Burleigh and Wolfgang Wippermann, *The Racial State: Germany, 1933–1945* (Cambridge: Cambridge University Press, 1993).

physician M.H. Jacobs responded to the common query as to whether a sample of human blood "could tell the race of the person from which it came." In most individual cases this was not possible, he explained, though some information could occasionally be obtained.

> We can, for example, in many cases say with a high degree of probability that a given sample of blood could not have come from a pure-blooded American Indian. It happens that in this race three of the four well-known blood groups, for which tests are made before blood transfusions, seem either to be absent or very rare, and consequently blood of any of these three types must, as a rule, come from persons belonging to other races.[520]

Jacobs was referring to the high frequency of blood type O—usually over 90 percent—found among Native Americans. Theirs was the only group surveyed that so closely suggested a direct correlation between blood group and racial type. In the vast majority of other peoples studied, including Germans, there had been no such obvious pattern; types AB and B were relatively rare, but both A and O were quite common. In Germany and elsewhere, there had been no reports of one blood type characterizing such a large segment of the population. Nevertheless, a select group of seroanthropologists continued to cling tightly to the idea that their science might one day assist anthropology.

Because of the ambiguity in seroanthropological research, expectations gradually increased during the Weimar Republic. Seroanthropologists could not research "just blood" but were forced to supplement their studies. To lend insight into blood type frequencies, many felt it was imperative to examine the specific settlement history of the subjects' native region(s). Thorough researchers also needed to compare their findings with those of similarly classed subject groups and locations. As Paul Steffan explained, if one examined the "indigenous peoples" of Bremen, they would also have to consider Ostfriesland. Similarly, seroanthropological research conducted in Hamburg, Holstein, or Herne would require analysis of Osnäbruck, while Freiburg's location in the Black Forest necessitated examining Peterstal.[521] Because it was commonly theorized that "racially pure"

---

[520]  M.H. Jacobs, "What Blood Tells," *Scientific Monthly* 36, no. 4 (1933): 367.
[521]  Steffan, ed., *Handbuch*, 398.

peoples could be found in isolated areas, it was necessary to gather statistics there to establish a comparison template that could then be used for analyses of larger towns and cities. The point was that serological distributions were not to be analyzed out of context. Researchers also had to be very careful who they examined. With the implication that such groups were not normal, and therefore not representative of the general population, the surveying of hospital and prison inmates, though convenient, was ultimately considered worthless for gauging a peoples' racial identity. Furthermore, once an appropriate subject group was selected, not only were individual blood samples to be taken, but it was also strongly encouraged that physiognomic characteristics be examined as well. A respectable seroanthropological study was expected to be very thorough. Still, even under ideal circumstances, the most meticulous blood type research rarely yielded a quick or even clear response in turn. This in and of itself would not have dissuaded Nazi bureaucrats and scientists from its study, but the implications of existing research likely did, as blood type distribution patterns among Germans frequently conflicted with *völkisch* ideals. Consistently high levels of blood type A, the type associated with Western European descent, were far from the norm. Even with isolated pockets of "native Germans" that were identified through serology, other results often suggested to examiners that substantial miscegenation with non-German peoples had occurred in the past. In addition to this explanation, which clearly conflicted with the racial principles propagated by the far right, blood science further linked groups that had long been recognized by anthropologists as racially distinct. On more than one occasion, subjects with very different historical backgrounds and physical features had precisely the same distributions of blood type: for instance, the Cantonese of China were serologically identical to the people of Katanga in the Congo and the Kazans of Russia. The inhabitants of Greenland even matched the aboriginal population of Australia.[522] As for the notion that blood type A was indicative of "Aryan descent," research also revealed that the Aborigines of Australia had levels of type A comparable to, and sometimes even higher than, the distributions reported among Germans—

---

[522] Schneider, "Chance and Social Setting," 560.

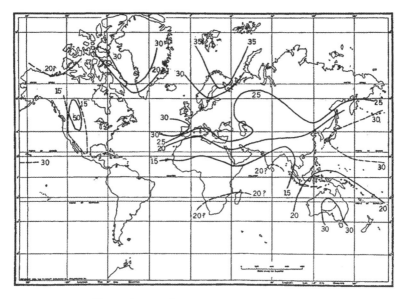

Figure 15. World map showing distribution percentages of type A blood. From
Leland Wyman and William Boyd, "Human Blood Groups and Anthropology,"
*American Anthropologist* 37, no. 2 (1935): 181–200.

thereby bringing into question the reliability of blood type A as an indication of "Aryan descent."[523] (See Figure 15.)

Blood had an additional drawback in that it could not be used to assist
in objective racial identification of Jews—the group with which the Nazis
were most concerned. No one blood type distribution, much less single
blood type, had been found to be uniquely Jewish. In fact, in seroanthropological studies Jews often closely resembled the majority population—
this, too, was anathema to Nazi race theory. In 1934 Nicolas Kossovitch
observed that certain Jews were found to have a percentage of types A and
B blood very similar to that of Europeans; German and Dutch Jews were,
respectively, similar to German and Dutch gentiles, while Persian Jews and
most of the Jews of the Middle East were comparable to the Asian peoples.
These matches led him to conclude that Jews served as one of the "best
examples" of changes in racial makeup due to miscegenation.[524]

---

[523] See Leland Wyman and William Boyd, "Human Blood Groups and Anthropology," *American Anthropologist*
37, no. 2 (1935): 189.

[524] Dujarric de la Riviere and Kossovitch, "Les groupes sanguins en anthropologie," 286.

Even blood type research aimed at individual racial diagnosis was not useful. Introduced in 1925, Manoiloff's unique chemical-based technique initially seemed promising because it differed from the Hirszfelds' broad, survey-based approach. Manoiloff claimed that he could tell the race of an individual from a sample of his or her blood alone. Since then, however, more extensive research indicated that the mysterious chemical reaction was unreliable. In the meantime, Manoiloff was forced to admit that he could not explain the chemistry behind his technique. One critic described Manoiloff's procedure as incorporating "various ingredients" in the manner of a cookbook recipe, in which "so much of this and so much of that is added, some kind of result is obtained, sometimes good and sometimes bad."[525] In response, Manoiloff could only weakly speculate that the blood of different races "probably had a different capacity of oxidation." In the larger discipline of seroanthropology, interest in Manoiloff's method was short-lived. The majority of blood type studies continued to rely upon the survey technique and, by the early 1930s, hundreds of thousands of individuals had had their blood analyzed for racial purposes. Despite the impressive numbers, there was not much to show for it. As one physician noted, "we still do not possess a method which enables the identification of an individual through their blood."[526] Such developments and the notable lack of progress explain the gradual decline in studies of blood and race. This shift is illustrated well in the opinion of race theorist Hans Günther. In 1922, after seroanthropology was first introduced, Günther acknowledged that there were "broad differences" between the blood of the different racial types and speculated that "extremely informative results" might be eventually be gleaned from blood science. His perspective had changed drastically by 1933, when Günther warned that it was clearly not possible to determine an individual's racial type from his or her blood (*die rassische Zugehörigket eines Menschen aus seinem Blut ablesen*).[527]

---

[525] Charles E. Abromavich, Jr., and W. Gardner Lynn, "Sex, Species, and Race Discrimination by Manoiloff's Methods," *Quarterly Review of Biology* 5, no. 1 (March 1930): 71. Despite his own criticism, Abromavich conceded that "nevertheless, the fact remains that under certain conditions correct results may apparently be obtained" (Ibid., 72).

[526] Paul Moureau, "Contribution a l'étude des facteurs d'individualisation du sang humain et leurs applications en medicine légale," *Revue Belge des Sciences Medicales* 7, no. 3 (1935): 178.

[527] Gerhard Baader, "Blutgruppenforschung im Nationalsozialismus," in Mariacarla Gadebusch Bondio, ed., *Blood in History and Blood Histories* (Florence: Sismelm, 2005), 342.

## Proponents of Seroanthropology

In spite of these drawbacks, a minority remained optimistic towards sero-anthropology and pointed to its contributions. One researcher commented that blood science allowed examiners to "isolate peoples of different origin living within the same country."[528] The "group dynamic" of seroanthropology, however, had been for many its most frustrating characteristic. To this, proponents of the science responded that blood was only one trait within a mosaic of racial indicators, and they would continue to maintain that there was a connection between appearance and blood type.

An article to this effect appeared in the 1933 edition of the *Zeitschrift für Rassenphysiologie* in which physician Hermann Gauch contrasted his subjects' blood types with "[Hans] Günther's markings of race"—the pigmentation of the eyes and hair, but also included other, "particularly obvious" racial traits, such as hair texture and brown eyes.[529] Reference to the work of Günther hints at Gauch's political tendencies, as does his labeling of blood types A and O as "Nordic" and B "non-Nordic." Based on their blood alone, which had already been typed, as the subjects were a selection of volunteers from a naval unit in Kiel, Gauch categorized 83 percent of the 250 men as Nordic.[530] Following a more detailed analysis encompassing a range of physiognomic traits, however, this could shift to a different category—usually Nordic, Eastern, or Dinarian. Gauch compiled a list including each subject's name, native region, racial type, most pronounced visible racial characteristics, and blood type:

> Thomsen, Schleswig, Nordic-Eastern, narrow eye-slit, broad cheekbones, blue eyes, wavy light-blond hair, type A
>
> Bitterlich, Saxony, Nordic, somewhat broad cheekbones, medium-blond straight hair, blue-gray eyes, type A
>
> Schmidt, Altona, Nordic, brown eyes, light blond straight hair, type A

---

[528] Dujarric de la Riviere and Kossovitch, "Les groupes sanguins en anthropologie," 287.

[529] Hermann Gauch, "Beitrag zum Zusammenhang zwischen Blutgruppe und Rasse," *Zeitschrift für Rassenphysiologie 6* (1933): 116–117. Gauch studied a selection of volunteers from a naval unit in Kiel, where the men's blood had been typed for transfusion records.

[530] Of 250 men studied, 35 percent had type O, 46 percent had type A, 14 percent had type B, and 5 percent had type AB. Ibid., 117. Gauch expressed an early interest in serology; his 1924 dissertation was on the blood types of "Dinarians" in Bavaria.

Rindfuss, Hesse, Nordic with some Eastern influence, broad cheekbones, grayish-brown eyes, black straight hair, type O [531]

Overall, the results supported Gauch's contention that blood types A and O were Nordic, as both were predominant among the participants who "looked Aryan." By contrast, he reported that those individuals with type B blood did not have *any* Nordic characteristics.[532] Gauch's final racial classifications belied such sweeping generalizations and claims of objectivity. In fact, the men's range of traits often prevented a single racial label. Gauch listed some individuals as "Nordic with Eastern influence," others "Falisch and Mediterranean." One subject was described as Nordic, Falisch, *and* "Cro-Magnon"—categories that would usually be considered mutually exclusive from one another. Predictably, an explanation of what distinguished the racial affiliation of one trait from the next was conspicuously absent, as was how each factored into Gauch's final decision, though he evidently felt it significant to note such traits as "a somewhat plump chin, a unibrow, freckles, and large nostrils and earlobes."[533] Gauch's research was an example of what was becoming, by this point, an increasingly rare attempt to demonstrate a correlation between blood type and physiognomy. The fact was that Gauch wanted there to be an association between blood and race, this association would serve as further proof that there were separate races. A Nordic bias was further evident in his general discussion of the field of seroanthropology, which, he inaccurately claimed, had been established by Paul Steffan. In reality, Steffan's work came several years after Hirszfeld's and, with its classification of type A blood as "Atlantic" and B as "Gondwanic," amounted to nothing more than a variation on the Hirszfelds' existing theory of "Western" type A and "Eastern" type B blood. This reference is all the more surprising when one considers that Gauch had completed his 1924 dissertation on seroanthropology. He must have been aware of the Hirszfelds' research and its critical role in the development of the science. It is also highly possible that Gauch, an anti-Semite—he advocated using blood science to "uncover Jewry's concealment" of its racial characteristics—was aware of Hirszfeld's Jewish background, which may have accounted for this omission.[534]

---

[531] Ibid.
[532] Ibid.
[533] Ibid., 118.
[534] Ibid., 116.

## Racial "Reform" under Nazism

The process of *Gleichschaltung*, in which various institutions were aligned with the National Socialist worldview, was not without problems in the case of racial anthropology. German race theorists were not a homogeneous group and, before 1933, were actually divided over the question of "Nordic supremacy." Some wanted nothing to do with the National Socialist variant of race theory. Others, such as Hans Günther, Alfred Ploetz, Fritz Lenz, and Ernst Rüdin, sympathized with Hitler. Differences between the two groups were regional, religious, and political.[535] During Weimar, seroanthropology had been similarly divided, between researchers with racial biases and those with a more liberal perspective. The large contingent of Jewish blood scientists fit into the latter group, though the authorship of seroanthropological studies in Germany reveals that the discipline overall had more appeal for the far right. Serologists of Jewish descent were generally less interested in racial applications of blood science than in its therapeutic and legal uses. By the early 1930s, frustrated by the relatively inconclusive research of the past decade, the group of unbiased blood scientists had largely diminished, along with those intent on exploiting the science for their own political purposes. Their work, as well as references to definitive racial types despite blatantly subjective evidence, was more prone to sweeping generalizations and vague terminology. It would seem, then, that blood science was in a unique position to benefit from the Third Reich and its emphasis on racial identification. Nazis regularly spoke of definite "racial types," even if they did so while exaggerating or blatantly misappropriating medical fact. The use of blood rhetoric, which had been so common in National Socialist propaganda and racial doctrine throughout the Weimar years, easily made the transition into the post-1933 period. One of its most conspicuous placements would be in the racial legislation adopted during the Third Reich.

The first attempt to enforce Nazi racial ideology through legislation came shortly after Hitler's appointment as chancellor, with the April Laws of 1933. References to blood were put to use not in the definitions them-

---

[535] Members of the Berlin Society for Racial Hygiene (Schallmayer, Baur, Ostermann, Muckermann, Grotjahn) tended to be more reluctant than their Munich colleagues (Ploetz, Gruber, Lenz, Rüdin) to embrace the fascist notion of an ideal Nordic type. See Proctor, *Racial Hygiene*, 46.

selves, which categorized individuals as "Aryan" or "non-Aryan," but in their interpretation. Everyone started with a clean slate as of January 1, 1800, and was assumed to be Aryan.[536] A party order concerning that date declared, "Prior to the emancipation of the Jews, the penetration of their blood into German folkdom was virtually impossible."[537] If no ancestor was either non-Aryan or the spouse or offspring of a non-Aryan after January 1, 1800, a person's "Aryan status" could not be altered by further genealogical exploration.[538] After that date, non-Aryan ancestry could alter one's status. Special concern was expressed with Jewish lineage; non-Aryan status applied "even if only one parent or grandparent was of Jewish faith."

If an individual's Aryan status was in question, an opinion was to be obtained from a "racial expert" (*Gutachter*) commissioned by the Reich Ministry of the Interior.[539] The problematic nature of the terminology in the April Laws quickly became apparent. Race ideologues had long employed the term "Aryan," traditionally used to identify a language group, to designate the Germans' racial superiority.[540] Throughout its years of misuse, others had on occasion called attention to this inaccuracy of racial discourse. German anthropologist Rudolf Virchow called the theory of an Aryan race "pure fiction."[541] French archaeologist Salomon Reinach referred to it as "prehistoric romance." German scholar Max Mueller bluntly rejected the distortion of his concept.[542] Use of the term "Aryan" in legislation prompted further disapproval, as well as confusion. Historian Saul K. Padover wrote that there was no such thing as an "Aryan race and ... the Germans are nei-

---

[536] Richard Lawrence Miller, *Nazi Justiz: Law of the Holocaust* (Westport, CT: Praeger Publishers, 1995), 15.

[537] Ibid.

[538] Ibid.

[539] Ibid., 13.

[540] "Hitler was undoubtedly unaware of the irony that the term 'Aryan' he so proudly bandied about had been introduced into popular usage by the son-in-law of Moses Mendelssohn, the eighteenth-century 'German Socrates,' who, as we have seen, interpreted Judaism in the spirit of the German Enlightenment and paved the way for Jewish intellectual participation in the wider society. Like Mendelssohn, his Christian son-in-law Friedrich Schlegel, a Romantic novelist, historian, and diplomat, campaigned for Jewish emancipation. But whereas previously the Germans, like all Christians, had sought a biblical origin for themselves (from Adam to Noah to Jepheth to Ashkenaz) and had speculated that German and all tongues derived somehow from Hebrew, studies carried out by Schlegel and others pointed, quite correctly, to a special relation of most ancient and modern European languages to Persian and Sanskrit." Frank Heynick, *Jews and Medicine*, 416.

[541] While investigating German heritage, Virchow conducted a census of hair and eye color of six million schoolchildren and examined skull deformities and other physical characteristics of ancient remains. Virchow concluded that modern Germans had inherited traits from multiple races. Later, in the 1930s, the Nazis tried to conceal his findings, which conflicted with their ideology of Aryan superiority and "pure German" lineage.

[542] Wallace R. Deuel, *People under Hitler* (New York: Harcourt Brace, 1942), 199.

ther Aryan, nor Nordic, nor a race."[543] Many ordinary Germans were uncertain of the terminology as well. "I looked up 'Aryan' in the dictionary, and it says they live in Asia," one citizen wrote to a sexton's office in Zehlendorf, a suburb of Berlin.[544] "There isn't any branch of my family there," the individual explained; "we all come from Prenzlau."[545]

While no explicit reference to blood was made in the laws' definition of "non-Aryan," it became common in the laws' discussion, application and interpretation. While making his case for a *Sippenamt* (Genealogical office) to assist in racial identification, Achim Gercke, an official in the Reich Ministry of the Interior, argued that such an office would "watch over the purity of the blood" (*Blutsreinheit*) of the *Volk* and further added that those who worked in the office would represent the "best German blood."[546] This type of ambiguity only exacerbated existing uncertainties. Largely because of the confusion in terms, state-appointed racial experts were inundated with requests. In December 1933, less than a year after the laws were introduced, correspondence within Württemburg reported the number of individuals in the Reich Health Council to be 181—a higher-than-average number, it was noted, "due to an increase in demands for racial experts," and requests for additional experts would continue until the end of the war.[547] In 1938 Berlin-based expert Dr. Otmar von Verschuer asked for more assistants, as he was unable to complete more than five requests per week.[548]

By early 1935, the ongoing queries forced the League of German Jurists to publish a summary of the problems related to the laws. Its author, Falk Ruttke, director of the Reich Committee for Public Health Service (*Volksgesundheitsdienst*), acknowledged the lack of clarity:

> While logic and consistency have traditionally been a special province of jurists and lawyers, it appears that since the *Machtergreifung* these faculties have eluded them. In reviewing our racial laws, it is apparent that these are lacking a certain conceptual clarity (*Begriffsklarheit*) in their use of such terms

---

[543] Saul K. Padover, "Who Are the Germans?" *Foreign Affairs* 13 (April 1935): 509–518. After receiving his PhD in history, Padover, an Austrian Jew who emigrated to the United States at fifteen years of age, worked at a number of posts in the U.S. Department of the Interior (1938–1944).

[544] Quoted in Deuel, *People under Hitler*, 195.

[545] Ibid.

[546] Hutton, *Race and the Third Reich*, 92.

[547] E 130b/2764.

[548] R 3001/20487, 12.

as "race," "racial hygiene," "eugenics," and others which fall into the same category. These are oftentimes used with different and even contradictory meanings.[549]

In spite of such difficulties, Nazi bureaucrats became even more insistent in their racial policies with the institution of the new, revised Nuremberg race laws approved by the Reichstag on September 15, 1935. This time, in order to prevent further confusion, the bureaucrats in the Ministry of the Interior who drafted the laws, "Jewish expert" Bernhard Lösener and state secretary Wilhelm Stuckart, chose to replace the previous category of "Aryan" with that of "German-blooded."

Blood rhetoric was central to the Nuremberg Laws and particularly to the anti-miscegenation law "for the protection of German blood and honor." It began with a preamble explaining that the Reichstag had unanimously adopted the law "imbued with the consciousness that the purity of German blood is essential to the continued existence of the German people."[550] The law forbade marriage and sexual intercourse between Jews and those of "German or related blood" (*deutschen oder artverwandten Blutes*). Other sections of the three-part legislation were similarly phrased. Reference was made to the "purity of the blood" (*die Reinheit des Blutes*), and the Citizenship Law restricted citizenship rights to those "of German or similar blood."[551] The German populace may have been more complacent about references to blood and race than about Aryanism, but the terminological shift did not help reduce the uncertainty encountered in the first set of race laws. Allusions to blood and race were nothing new, but their inclusion in actual legislation was a considerable divergence from past use in propaganda. As a result, the laws often had the opposite effect of that intended, and uncertainties regarding who was, and who was not, "German-blooded" quickly followed.

As was the case previously, commentaries and supplements on the laws needed to be issued, though these too were rife with ambiguity. In its simplest form, the rule defined a Jew as someone descended from at least three

---

[549] Falk Ruttke, "Erb- und Rassenpflege in Gesetzgebung und Rechtssprechung des Dritten Reiches," *Juristisches Wochenschrift* (May 11, 1935): 25–27.

[550] The laws were printed and distributed the following day. Joseph Tenenbaum, *Race and Reich: The Story of an Epoch* (Westport, CT: Greenwood Press, 1976), 3.

[551] Miller *Nazi Justiz*, 221–228.

Jewish grandparents. For the many thousands of individuals of mixed German-Jewish lineage, however, classification was contingent upon the proportion of "Jewish blood." At the extreme end were Nazis like the notoriously disreputable Julius Streicher, who had proposed that "one drop of Jewish blood" was sufficient to exclude a person from the Aryan race.[552] Instead, officials explained that all *Mischlinge* (individuals of "mixed" lineage) with 75 percent or more Jewish blood were to count as Jews; on the other hand, not necessarily all German-Jewish *Mischlinge* with less than 75 percent Jewish blood were legally Jewish.[553] If a *Mischling* belonged to the Jewish faith, however, he or she would then be classified as a full Jew. Blood was still a malleable concept, as it had been all along—race was contained in the blood, but it could also be affected by an individual's religion. This flexibility complicated effective interpretation, and therefore enforcement, of the laws. Even National Socialist bureaucrats found the phrase "German-blooded" awkward. In December 1935, shortly after the laws were introduced, the Reich Ministry of Justice distributed an interdepartmental memo that again outlined the criteria for "proof" that an individual was "German blooded" (*Nachweis der deutschblütigen Abstammung*):

1. The individual's grandparents are first considered when determining Aryan descent. Their relation is confirmed through civil or official parish documentation.

2. It might then be established that none of the grandparents are of the Aryan race. As non-Aryans, it is then considered whether or not they are members of the Jewish faith.

3. To prove that the individual's grandparents were never practicing Jews, the birth certificates of the grandparents must be provided. Official parish confirmation is also sufficient—particularly in the event that the grandparents were baptized.

4. In some instances, Aryan descent cannot be determined due to missing information or documentation. In such cases, an expert from the Reich Office for Genealogy (*Reichstelle für Sippenforschung*) in Berlin is to be contacted.[554]

---

[552] Karl A. Schleunes, ed., *Legislating the Holocaust: The Bernhard Loesener Memoirs and Supporting Documents* (Boulder, CO: Westview Press, 2001), 9.

[553] R1501/5513, 17.

[554] R3001/24433, 6–7. From Dr. Freisler, Reich Ministry of Justice, December 7, 1935, to: Herrn Präsidenten des Reichsgerichts, Herrn Oberreichsanwalt, Herrn Präsidenten des Volksgerichtshofs, Präsidenten des Reichspatentamts, Präsidenten des Landeserbhofgerichts in Celle, Oberlandesgerichtspräsidenten, Generalstaatsanwälte bei den Oberlandesgerichten.

Issued by Roland Freisler, these specifications came "from the top." Freisler was a dedicated Nazi who served as a ministerial director in the Prussian Ministry of Justice. He was also the most famous judge in the People's Court system, notorious for his tirades, degrading of defendants, and ruthless sentencing.[555] In this interpretation of the Nuremberg Laws, Freisler reverted back to the term "Aryan" instead of "German-blooded," thereby indicating that the two were essentially interchangeable. The requirement for definitive proof of either "Aryan or German blood" is still evident in the heavy reliance upon documentation and, again, the importance of religion in determining what was portrayed as a biological category is clear. As it had been with the April Laws, "racial experts" were to be consulted in the event that documentation was lacking.

Without the preferred documentation, racial experts would usually refer to physiognomic characteristics to determine the racial type of the subject in question. The examination would most often consist of a blood test, as well as scrutiny of the shape and physiology of the eye and the shape of the head, along with photographing from the front and in profile.[556] These were general patterns, however, and the characteristics chosen, as well as their individual significance in determining race, could vary considerably from one investigator to the next. Because there were no absolute guidelines, examiners were able to exercise their individual preferences. If a blood sample was drawn, it was usually done to test for disease and was not part of the racial analysis. Blood was not mentioned by Nazi officials in the range of criteria recommended to ensure accurate racial categorization. Also to be taken into account were the individual's "disposition and mental state." Even then, however, physiognomy trumped all other indicators, as descendants of Eastern Jews or even half-Jews with a "strong Jewish appearance" were to be declined (categorized as non-German blooded). At this point, *Mischlinge* could then apply for "voluntary sterilization."[557]

As in past anthropological studies, classification of *Mischlinge* was often problematic. In a letter to the Reich Ministry of the Interior, Bernhard Lösener commented that it was very difficult, and sometimes not even

---

[555] Michael C. Thomsett, *The German Opposition to Hitler: The Resistance, the Underground, and Assassination Plots, 1938–1945* (Jefferson, NC: McFarland, 2007), 83.

[556] Schafft, *From Racism to Genocide*, 73.

[557] R1501/5513, 45.

possible, to racially categorize an individual with "exactly 50 percent Germanic and 50 percent Jewish blood."[558] In these cases, he explained that a decision had to be made by analyzing their external appearance and other factors (such as family history or the individual's economic and political perspectives). However, considering Lösener's additional claim that each "half-Jew" could possess "thousands of Jewish and German characteristics," this would not necessarily be helpful.[559] Although his colleagues and racial experts themselves focused on appearance, Lösener believed that the most important characteristics in determining race were the "unseen traits." In the end, even physiognomy was not always foolproof, and determining serological type, despite the prospect of "German blood," was never a priority of racial experts. Given the inconsistent guidance, and the extensive list of potential racial indicators, even Nazi bureaucrats acknowledged that expert analysis could result in inaccurate racial categorization.[560]

## "Blood Defilement"

The Nuremberg Laws indicated the Nazis' primary racial objective of keeping the blood of different races separate and were directed most strongly against the perceived "influence of Jewish blood" (*jüdisches Blutein-schlages*). The malleability of blood-related propaganda allowed for virtually endless interpretations as to how exactly "Jewish blood" was a threat. In a *Ziel und Weg* article on the "solution of the Jewish Question," Alfred Böttcher explained that the Jews had never "had a fatherland" or built their own state; the reason for this was an absolute "destructive power" in their blood.[561] Others proposed that Jews actually needed to acquire non-Jewish blood for religious ritual. The Nazi tabloid *Der Stürmer* reported how, in Madrid, the body of a four-year-old child had been found with two deep wounds in its throat, drained of blood, and went on to explain how this was believed to have been the work of the Jews and their centuries-old "vampire secrets" (*Vampir-Geheimnissen*).[562] Propaganda on Jews and blood most frequently centered on "blood defilement," which was most often

---

[558] Ibid., 141–142.
[559] Ibid.
[560] R1501/5513, 142.
[561] Alfred Böttcher, "Die Losung der Judenfrage," *Ziel und Weg* 5 (1935): 226.
[562] "Die Sünde wider das Blut," *Der Stürmer* 2, no. 7 (May 1924): 2.

depicted as the result of miscegenation between Jews and non-Jews. Nazi ideologues repeatedly emphasized the adverse effects of blood mixing and the necessity of guarding the purity of one's own blood. One *Der Stürmer* article, entitled "The Sin against the Blood," pointed to the New World as one example of the disadvantageous consequences of miscegenation:

> The discovery of the Americas has also had many unfortunate consequences. The Spanish, who settled especially in South and Central America, mixed over the centuries with the natives of the land. This resulted in a mixed race [*Mischvölker*] in which the blood of the Indians and Negroes [*Negerblut*] was now also carried by Europeans.[563]

*Mischlinge*, the account continued, "would always have the worst characteristics of both parents." They were not capable of being either creative or cultured—it was the "sin," their mixed blood, that prevented them from ever living in peace or being truly productive. The reader had only to look to the "bastard island of Cuba" or the Balkans—both areas of persistent political unrest—for further confirmation of this phenomenon.

In a 1934 interview, prior to the Nuremberg Laws, German race theorist Eugen Fischer explained that women were central to effective eugenics to prevent "damage to our [German] blood." Men were accountable as well, as an error in judgment could cause an "unfortunate creature," or "bastard," to be born, who would then be afflicted with two "opposing bloods" (*Blutarten*) at war within them."[564] The Nazis expressed more concern over women's sexual choices, with the understanding that they were uniquely physically susceptible to blood defilement through intercourse. Some Nazis explained that the semen of a Jewish man was absorbed into the blood of an "Aryan woman" during intercourse, thereby infecting her and permanently altering her racial identity. As a result, even one sexual encounter with a Jew was believed to cause sufficient, *permanent* damage. (Despite Fischer's warning, the same concern did not apply to German men, who, it was thought, could often have sex with Jewish women without similar consequence.) During the Third Reich, this belief led to a closer policing of relationships between Aryan women and Jewish men. The idea of sexual blood

---

[563] Ibid.

[564] Charlotte Köhn-Behrens, *Was ist Rasse? Gespräche mit den grössten deutschen Forschern der Gegenwart* (Munich: Eher Verlag, 1934), 51–52.

defilement, which had been propagated for years in extremist tracts such as Artur Dinter's *The Sin against the Blood*, was incorporated for the first time into actual legislation in the Nuremberg Laws. To stop such "bloodletting," the legislation forbade sexual contact between what were considered incompatible racial types.[565] Marriages were only permitted if they did not endanger the purity of German blood, and sexual activity with "racial others" was penalized accordingly.[566] Which bloods could legally "mix" without threatening German blood led to a complicated series of restrictions on marriages between Jews and Germans, Jews and quarter-Jews, Jews and still-uncategorized half-Jews, Germans and still-uncategorized half-Jews, half-Jews with each other, and half Jews and quarter-Jews.[567]

In practice, the circumstances that resulted in defilement of blood were debatable. Even with repeated caution that miscegenation between a Jewish man and a German woman would permanently pollute the blood of the racially superior (female) participant and her children, a statement issued by Dr. Arthur Gütt, the ministerial director of the Reich Health Office in the Reich Ministry of the Interior suggested that the results were not irreversible:

> It is possible that the mixing of a quarter and a half-Jew with an individual of German blood could result in the complete elimination of Jewish characteristics. In the event that no further crossing with Jews would take place, it can be assumed that any existing Jewish blood would gradually be diluted over generations. After many generations, one comes progressively closer to a pure Aryan. Under these conditions, the absorption (*Aufsaugung*) of a Jewish *Mischling* into the German people would be possible.[568]

With special permission, even marriages between "half-Jews" and those of German blood could be granted.[569] These conflicting descriptions raised the question: could Jewish blood be "diluted" over time, or was it permanent? National Socialist references to seroanthropology were very few, even though certain studies were applicable to the Nazis' concerns and might have been useful in justifying state policy.

---

[565] Schafft, *From Racism to Genocide*, 217.
[566] R1501/5513, 26–27.
[567] Ibid., 143.
[568] Ibid., 43.
[569] Ibid., 44.

From one perspective, the research of Russian seroanthropologist E.O. Manoiloff had demonstrated the "adverse affects" of miscegenation. Manoiloff suspected that his technique of racial identification, which relied upon the color of a blood sample after exposure to certain chemicals, might be helpful in cases of disputed paternity in which a child was "racially mixed." Upon examination, he reported that the blood of a child born of a "pure Russian father" and "pure Jewish mother" yielded a color different from that of the Russian alone. A similar result was obtained in a separate instance involving a Russian father and German mother. In these cases, Manoiloff noted that the reaction color of the child's blood, which represented the child's racial type, matched neither parent. Even if a subject was "racially pure" (Russian, German, or Jewish) in appearance, his or her blood might reveal otherwise. Manoiloff's technique was seriously flawed, however. Results were subject to variations based on sex, race, and how closely related the racial types of the parents were. If the father was Russian and the mother Jewish, Armenian, or Polish, the reaction of the offspring's blood more closely resembled that of the mother. By contrast, if the father was Russian and the mother German, Finnish, or Tatar, the color differed very little from that of an "unmixed Russian."[570] Nonetheless, Manoiloff still claimed that his reaction could have a certain "medico-legal importance if it relates to the problem of determining the parentage of a child born of a Russian mother and a Jewish father."[571] In theory, had it been consistent, Manoiloff's work would have been useful in cases of disputed paternity. Due to its unreliability, however, researchers were understandably hesitant to apply Manoiloff's chemical reaction to either racial or paternal cases, and it instead met largely with disregard or outright criticism.[572]

---

[570] Poliakowa, "Manoiloff's 'Race' Reaction," 25.

[571] Manoiloff, "Discernment of Human Races by Blood," 20.

[572] The first attempt to determine paternity based on "biometric methods and biochemical reactions" was proposed by Russian professor Poliakowa, and his analysis involved 125 questions. Poliakowa was intrigued by the prospect of a more efficient means of resolving disputed parentage. Prior to examining the legal ramifications of Manoiloff's methodology, she first tested its reliability in determining race. She studied the reaction colors of blood samples taken from Russians, Jews, Estonians, Lithuanians, Koreans, Poles, and Kyrgyz. Her findings suggested the accuracy of Manoiloff's work, as the blood of each group had turned a characteristic color. Evidently satisfied with the results, she later decided to publish the results of a study in which she had used the serological method of chemical analysis to settle a case of disputed paternity. Nonetheless, this method of classification never progressed in American courts. Poliakowa's apparent "success" was an isolated event. See Poliakowa, "Manoiloff's 'Race' Reaction," 23–29.

National Socialists never made clinical use of Manoiloff's findings, but they were given fleeting reference in the virulently anti-Semitic *Der Stürmer*.[573] Although over a decade had passed since Manoiloff published his study, and his obscure technique had since met with complete rejection from the medical community, the article "Blood and Race" excitedly detailed his findings:

> Professor E.O. Manoiloff, a Russian biologist, has been successful in differentiating the blood of Jews from non-Jews with the aid of [catalyst] cresyl-violet. In the blood drawn from Jews, the color of the cresyl-violet disappeared, whereas in the blood of the non-Jews the color remained partially the same and took on a bluish-red hue.

Of course, the article did not mention the inconsistencies in Manoiloff's work, the fact that one often had to interpret shades of color or the misgivings of racial examiners—all of which had resulted in the obscurity of the serochemical reaction by the time of this report. The article continued nevertheless, explaining how the 1927 research of an American "Ms. [Doctor] Poliakowa" had recorded the blood of *Mischlinge* changing to the color indicative of a "racially pure Jew." In *Der Stürmer*, this was sufficient proof that the children of mixed marriages inherited the characteristics of the "racially inferior" parent.

Despite the optimism of the glowing review of *Der Stürmer*, Manoiloff's technique was not referred to either by German race scientists for systematic racial classification or by German lawyers for paternity testing. Only the actual blood types could be used as legal evidence in excluding a putative father, and reference was never made linking blood type to the race of the person in question, though Otto Reche maintained that "in certain instances, it can be determined whether or not the father of an illegitimate child is Jewish, because Asian type B blood is more common among Jews than among Europeans."[574] Reche's opinion was not shared by legal authorities; type B blood was not referred to as Asian or Jewish in court documents. Even in his own practice, Reche explained that blood typing only allowed one to determine whether paternity was even a possibility.[575]

---

[573] "Blut und Rasse," *Der Stürmer* 14, no. 31 (July 1936): 3.

[574] Köhn-Behrens, *Was ist Rasse?* 98–101.

[575] Otto Reche, "Zur Geschichte des biologischen Abstammungsnachweises in Deutschland," *Volk und Rasse* 13 (1938): 374.

This was definite progress over some previous methods. As one physician explained in 1899, the presumed father of an illegitimate child was the man who had lived with the mother "shortly before the birth of a child."[576] By the early twentieth century, determining who the biological father was was based on who had had intercourse with the mother from 266 to 282 days prior to the birth of the child.[577] Critics of this technique called attention to its obvious drawbacks, among them the fact that the length of a pregnancy could vary considerably from one individual to the next. Blood types presented an "obvious advantage," Reche explained, but it was also necessary to analyze other dominant hereditary characteristics in order to attain definitive results. As with racial anthropology, reference to physiognomy was the oldest methodology, and Reche pointed out that this was occasionally made easier by cases in which the child and putative father shared physical abnormalities, such as "white hair, six fingers, or a hare lip." He further noted that cases in which the father belonged to a "completely foreign race" (e.g., "Negroes") were relatively easy to decide.[578] In the majority of cases, he concluded, it was necessary to simultaneously consider a range of inherited characteristics.[579] In spite of his own advice, Reche placed little emphasis on blood typing in determining parentage. Regardless of technique, whether it was Manoiloff's reaction or simply typing blood, no test had been able to identify "German blood." This had been proven in both medical and legal practice, and continued to be a source of confusion in what it meant to have "German blood." Clarifications from supposed racial experts were not necessarily helpful.

In his 1934 introduction to genetics and eugenics, German anthropologist Otto Rabes described the German people as only "50 percent Nordic"; the remaining half was a mix of various European racial types.[580] In a 1936 commentary on the Nuremberg Laws, senior German legal bureaucrat Ernst Brandis similarly explained that the Germans were not a "unitary race, but

---

[576] K. Albrecht and D. Schultheiss, "Proof of Paternity: Historical Reflections on an Andrological-forensic Challenge," *Andrologia* 36 (2004): 33.

[577] Geisenhainer, *Rasse ist Schicksal*, 125.

[578] Ibid., 370.

[579] Ibid., 374. As with his work with blood in race science, Reche was also constantly trying to identify a definitive "paternal indicator." In 1938 he wrote that he had already "for years" been examining fingerprint patterns between children and their fathers, which, to him, "seemed promising." Ibid., 372.

[580] Otto Rabes, *Vererbung und Rassenpflege. Versuche und Stoff für den Unterricht und Rassenpflege* (Leipzig: Quelle und Meyer, 1934), 44.

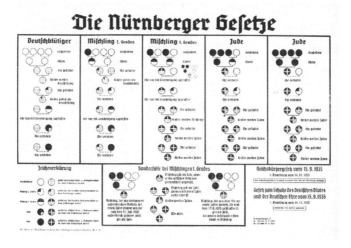

Figure 16. Chart for the Nuremberg Laws, from the Reich Ministry of the Interior, showing the proportion of "German" or "Jewish" blood that determined an individual's racial classification. From Annette Weber, "Blood as the Object of Scientific Discovery and Romantic Mystification," in James M. Bradburne, ed., *Blood: Art, Power, Politics, and Pathology* (New York: Prestel, 2002), 164.

rather one composed of various groups (Nordic, Phalian, Dinaric, Alpine, Mediterranean, East-Elbian) and mixtures between these." The "blood of all of these races and their mixtures," he explained, was that found in the German people, and therefore that "which represented German blood."[581] Similar statements were even made by officials at the Reich Ministry of the Interior. One document noted how, "over the course of millennia," Germans had encountered and absorbed other "unwanted bloods" (for instance, those of the Eastern and East Baltic races). This mixing was so minimal (listed specifically as ratios of ½ to 1,000, or ¼ to 1,000), the author declared, that any dangers to the (German) race could be disregarded.[582] Awkward, frequently inconsistent interpretations of the German race and blood were common then not only in scientific literature, but also among the Nazi bureaucracy. Because state authorities were not entirely certain of the Germans' racial classification, it was difficult for them to provide a logical explanation for the phrase "German-blooded." (See Figure 16.)

---

[581] Ernst Brandis, *Die Ehegesetze von 1935* (Berlin: Verlag für Standesamtswesen, 1936), 33.
[582] R1501/5513, 154.

Inquiries from the German public relating to the state's use of racial terminology persisted.[583] The volume of questions prompted the authors of the Nuremberg Laws to issue a published statement less than a year after the laws were passed. Under the impression that the "opponents of race theory would attempt to confuse the German people and prevent its acceptance altogether," Wilhelm Stuckart and Hans Globke felt compelled to do so.[584] Their commentary began by reiterating some of the most basic questions raised: "What do race, people, and state have to do with one another? What is a race? What is a *Volk*?" These terms were not interchangeable, despite their relatively fluid use in racial propaganda, Stuckart and Globke explained. Nationality (a *Volk*) was rooted in "constitutional law, national character, history, and culture," while race was a separate, "*scientific* notion." To lend further insight, the authors referred to the definition of race according to Hans F.K. Günther: "a race is a group of people who can be consistently differentiated from others by distinct physical and psychological characteristics." The pamphlet neglected to list the specific traits, or mention that opinions on which were the most important varied widely within the discipline of racial anthropology and also among the so-called "racial experts" authorized by the state to assist in racial identification.

## DIVERSE MEANS OF "BLOOD DEFILEMENT"

"Blood defilement" was most often depicted in a sexual context, though the Nazis believed that it could also result from other types of contact.[585] Decisions passed in certain Nuremberg court cases show that the "purity" of an individual's blood could be affected even without intercourse.[586] The

[583] Karl A. Schleunes, *The Twisted Road to Auschwitz: Nazi Policy toward German Jews, 1933-1939* (Champaign, IL: University of Illinois Press, 1990): 115. For confusion relating to laws, see also Michael Robert Marrus, *The Nuremberg War Crimes Trial, 1945-1946* (New York: Macmillan, 1997); A. Margaliot, "The Reaction of the Jewish Public in Germany to the Nuremberg Laws," *Yad Vashem Studies* 12 (1977), 75–107; Robert Gellately, *Backing Hitler: Consent and Coercion in Nazi Germany* (Oxford: Oxford University Press, 2002), 122; David Cesarani and Sarah Kavanaugh, eds., *Holocaust: Critical Concepts in Historical Studies* (London: Routledge, 2004).

[584] Wilhelm Stuckart and Hans Globke, *Kommentare zur deutschen Rassengesetzgebung* (Munich and Berlin: C.H.B. Verlag, 1936), 1.

[585] "Even platonic relationships between a Jewish man and a German-blooded woman were subject to scrutiny and at times judged sufficiently close to result in a conviction of 'race shame' (*Rassenschande*). Patricia Szobar, "Telling Sexual Stories in the Nazi Courts of Law: Race Defilement in Germany, 1933 to 1945," *Journal of the History of Sexuality* 11, nos. 1–2 (2002): 131–163.

[586] R1501/5513, 27.

Nazis further believed that there were race-altering consequences from literal blood mixing in the form of transfusions. Racial ideologues contended that a blood transfusion could transform "an Aryan into a non-Aryan."[587] During the Third Reich, this notion was first officially indicated in an incident in late September 1935 in which a German Jewish physician donated his own blood to save the life of a member of the Stormtroopers (*Sturmabteilung*, or SA). The transfusion occurred in the town of Niederlungwitz, near Chemnitz, in the eastern German region of Saxony. After being critically injured in an auto accident, the SA man was rushed to a nearby Jewish hospital, where he was treated by physician Hans Serelman. At this point, stored blood was not commonly available in Germany.[588] If a transfusion was needed, a so-called "donor-on-the-hoof" was contacted—a volunteer who lived near the medical establishment. These individuals, whose blood had already been typed, were expected to be accessible in emergencies in which their serological type was compatible with that of the patient. With no such available donors, the Jewish Dr. Serelman decided to give his own blood, whose type happened to be compatible with that of the SA man. The man survived and fully recovered.

German physicians sympathetic to National Socialism subsequently denounced Serelman to the Gestapo for having injected Jewish, or "foreign," blood into a German patient.[589] An SA tribunal met to consider whether the receipt of this blood meant that the man should be expelled from the organization. Under state examination, Serelman admitted that he had donated his own blood to an "Aryan", and that this was not the first time that he had done so in an emergency (all of the patients so treated had regained their health). Because Serelman was a veteran of World War I, the court decided that his blood had not compromised the SA man's racial status or purity.[590] Their decision was made in accordance with the Law for the Restoration of the Professional Civil Service. Passed in April 1933 on

---

[587] Deuel, *People under Hitler*, 212.

[588] In the mid-1930s, blood transfusions in Germany were considerably less common and well-coordinated than those practiced in Great Britain and the United States. The Serelman case occurred sixteen years after Karl Landsteiner organized the first postwar American blood bank in 1921. Furthermore, it was two years after the Serelman case that the first full-scale blood bank for transfusion was set up with regular typing of blood in Germany. See Schneider, "Chance and Social Setting in the Application of the Discovery of Blood Groups," 546.

[589] Deuel, *People under Hitler*, 213.

[590] Ibid.

the insistence of President Hindenburg, this law held that German Jews who had fought in the Great War were to be exempt from National Socialist anti-Semitic legislation. The SA tribunal did explain, however, that if the Jewish donor had not been covered by the Hindenburg Exception, the SA man's racial classification would have been compromised.[591]

In spite of his veteran status and the fact that he had certainly saved numerous lives, Serelman was still charged with the crime of "race defilement" and sent to a concentration camp. Shortly after the judgment, a *New York Times* article noted the incongruity between the verdict and a statement made by Professor Leffler, a high functionary in the racial-political bureau of the Nazi Party. In spite of the court's decision, Leffler claimed that a change in racial character following a donation was "sheer nonsense." This notion, he explained, was simply the result of confusion derived from the *figurative* use of the word "blood" in regards to heredity. Blood donor cards issued during the Third Reich reflected Leffler's pragmatism. While individuals were required to confirm that they did not have "non-Aryan" blood, there were no medical criteria listed that differentiated between this and "Aryan" blood. Forms included the person's age, address, blood type, blood bank, date of donation, amount taken, and the results of hemoglobin and syphilis testing.[592] No blood types were excluded from donating. Type B, for instance, which had been repeatedly associated in seroanthropological study with Eastern, or "inferior" racial types, was not marked "non-German." Physicians were careful to label the blood group of each donation, not racial type, to ensure transfusion compatibility. By the mid-1930s, one German physician remarked that this necessity was thoroughly recognized by the medical establishment, as type incompatibility could result in death.[593] The blurred line between medical fact and Nazi race theory, exemplified by the Serelman case and the prohibition of blood transfusions between "Aryans" and "non-Aryans", did not go entirely unnoticed by the German public.

In February 1937, *Münchener Medizinische Wochenschrift* (Munich Medical Weekly) addressed whether blood actually differed, as one reader

---

[591] "Says Transfusion Can't Alter Race: Nazi Expert Holds Recipient of Alien Blood Will Not Be Affected by Operation; Leffler's Declaration Inspired by Case Involving a Jewish Doctor and 'Aryan' Patient," *New York Times* (October 20, 1935): 28.

[592] E151/54, 378.

[593] Oswald Streng, "Die Bluteigenschaften (Blutgruppen) der Völker," 407.

inquired, "from race to race."[594] Two physicians responded. The first, Albert Harrasser, explained that "no certainties" had been recognized in the racial differentiation of blood, though many attempts had been made to determine individual races through the biochemical index (the ratio that compared the incidence of type A to type B blood within a group). There was the possibility, he thought, that species-specific proteins characterized the different racial types. For further information on the matter, Harrasser referred to the *Zeitschrift für Rassenphysiologie* and Paul Steffan's edited volume *Handbuch der Blutgruppenkunde*. The response of Harrasser's colleague, Theodor Mollison, was much more decisive. Mollison expressed "no doubt" that there were differences between the blood proteins of racially disparate individuals, though he immediately added that the exact nature of this relationship was not entirely understood—nor had it been discerned among humans. For proof, Mollison cited the work of Dr. Carl Bruck, which had involved primates.[595] Nonetheless, Mollison explained that the Anthropological Institute of the University of Munich was planning to research this relationship. The discrepancy must have been painfully obvious to the reader: Mollison declared that there was a link between blood and race while simultaneously admitting that it had not yet been confirmed. Nevertheless, interracial blood transfusions were restricted in Germany, and after the *Anschluss*, Austrian hospitals took it upon themselves to use only blood certified as having come from Gentiles for transfusions performed on "German-blooded" patients, and only "Jewish blood" for those performed on "persons of alien blood."[596]

## SEROANTHROPOLOGICAL RESEARCH IN THE THIRD REICH

There were plans for other types of seroanthropological research, but in the 1930s, the traditional means of study remained the mass blood type survey. Interest in this approach remained low and was likely influenced by the state's need for individual racial classification, as well as the fact that it often turned up results incompatible with Nazi race theory. In his 1936

---

[594] Albert Harrasser and Theodor Mollison, "Fragen: Frage 27," *Münchener Medizinische Wochenschrift* (February 2, 1937): 265.

[595] C. Bruck, "Die biologische Differenzierung von Affenarten und menschlichen Rassen durch spezifische Blutreaktion," *Berlin Klinischer Wochenschrift* 44, no. 26 (1907): 793–797.

[596] Deuel, *People under Hitler*, 213.

study on blood group distributions, particularly those of Germans, Norwegian physician Oswald Streng remarked that "Slavic mixings" were apparent well into Western Europe. Furthermore, he claimed, the "invasion" of type B from Asia had been extensive, stretching back to the time of Genghis Khan and perhaps even earlier. The blood type surveys revealed that there had been repeated influxes of type B into Europe over the centuries. The distance between east and west explained why type B was much less pronounced in the northwest; Streng reported that it was least common in Spain, France, Belgium, Holland, England, and Scandinavia.[597] Because of its higher proportion of "Asian" type B, Germany was not included on this list. According to Streng, the possibility even existed that the German people were Asiatic in origin.[598]

Even though his research offered nothing qualitatively new in the way of racial analysis, Streng advocated additional blood type surveys in the hope that this might still offer certain benefits over existing methodologies. He noted the ease of determining both blood type and the racial biochemical index, as well as the fact that blood type was inherited according to certain strict principles—which presented an advantage over the "gradual nuances and combinations" of other anthropological characteristics, such as skull length, cranial index, and eye color.[599] Furthermore, he observed, blood type was an unchanging characteristic, a fact that was first confirmed by Verzár and Weszeczky, whose 1921 study referenced similarities between (German and Roma-Sinti) immigrants and natives of their respective homelands. Based on the blood type patterns in Germany itself, specifically the higher incidence of type B blood along its eastern border, Streng divided the German people into "Western" and "Eastern" types. He then compared these frequencies to the blood of ethnic German settlements in Hungary, Russia, and Romania. The type patterns of Germans in the East clearly resembled one another.[600] Streng observed that these groups of ethnic Germans were more similar in their blood types to western Germans

[597] Streng "Die Bluteigenschaften (Blutgruppen) der Völker," 425.

[598] Ibid., 429.

[599] Ibid., 408.

[600] German settlers type O 37.7 percent, type A 47.5 percent, type B 11.3 percent; West Germans type O 40.9 percent, type A 44.6 percent, type B 10.5 percent; East Germans type O 37 percent, type A 42.1 percent, type B 14.9 percent. Ibid.

than eastern Germans in Germany.[601] However, Streng put an abrupt end to any developing hypothesis with the simple statement that "all modern peoples are [racially] mixed" (*gemischte*), and this was followed by "it is difficult to say with any certainty from where they came."[602] Both quotes made his study not only irrelevant to the racial concerns of National Socialism, but unwelcome as well. His opening reference to the benefits of seroanthropology aside, Streng plainly rejected the notion that blood could racially discern between one group of people and another.

Siegmund Wellisch, who had been a devotee of seroanthropology during the Weimar Republic and an active member of the German Institute for Blood Group Research, concurred with Streng in an article published the following year. "Today," Wellisch remarked, "there are only combinations of different groups, and no longer any unmixed races. As a result, racial characteristics are neither qualitatively nor quantitatively apparent through observations and measurements."[603] Wellisch believed that the probability of singling out "racially pure" individuals through a homozygous group of characteristics was minimal—so low that it would be safe to say that there were hardly any pure races left in modern Europe. Because of miscegenation, Wellisch explained that individuals with "questionable racial identities" were common within every group.[604]

Wellisch believed that this principle applied to Jews as well.[605] Based on their extensive past movements and interactions with other peoples, Wellisch concluded that Jews had become thoroughly racially mixed even before their expulsion from Israel. Miscegenation continued as the Jews dispersed throughout the Mediterranean Basin and Europe.[606] It was only after

---

[601]  Streng, "Die Bluteigenschaften (Blutgruppen) der Völker," 411.

[602]  Ibid., 421.

[603]  Siegmund Wellisch, "Serologische Rassenanalyse," *Zeitschrift für Rassenphysiologie* 9 (1937): 38.

[604]  On the other hand, Wellisch continued, it was not uncommon to encounter a person with numerous characteristics associated with a distinct racial type—perhaps so many, he explained, that the person might be considered "representative" of that type. Therefore, it would on occasion be possible to recognize an "original, completely pure" race in some modern peoples but, Wellisch was careful to explain, this did not mean that it was absolutely necessary for a person to possess these same characteristics in order to belong to that same race. Ibid., 37.

[605]  Siegmund Wellisch, "Rassendiagnose der Juden," *Anthropos* 32 (1937): 783–794.

[606]  Wellisch traced the Jews' racial makeup as far back as 1700 BC. Over the course of their extensive history, he claimed, they had been racially affected by interacting with the following groups: the Amorites, Hittites, Mitanni, and Kuschiten (who were varying degrees of "Oriental, Nordic, Mediterranean, and/or Negro" descent). This "mixing" continued over the centuries with other groups—the "racially similar" Canaanites, the Phoenicians, Jebusites, Hivites, Gibeonites, and other Biblical peoples. Eventually, Wellisch explained, the

"centuries of isolation" that Jews separated into the modern Ashkenazi and Sephardi racial types, who were both "physiognomically and serologically distinct from one another."[607] This seemed to be a promising claim, particularly if one wanted to distinguish Jews from the larger population. However, just as Streng did, Wellisch followed what seemed to be useful advice with a caveat or comment that negated any previous case made for the usefulness of seroanthropology. The vast majority of Jews looked Ashkenazi, or so Wellisch reasoned, and their blood type distributions indicated an Ashkenazi connection as well. However—and this was a critical drawback in a state that placed the greatest emphasis on appearance in classifying racial types—there was no connection between physiognomy and blood type. This was made even worse by Wellisch's reference to the fact that all efforts to trace Jewish descent through blood science had been "futile."

## THE GERMAN INSTITUTE FOR BLOOD GROUP RESEARCH

Even though seroanthropological studies were at times unfavorable towards, and even incompatible with the Nazi racial worldview, the German Institute for Blood Group Research continued its pursuit of the science and the political agenda of its administrators—who were sympathetic to Nazi beliefs since the institute's establishment in 1926—seemed not to have changed. Notwithstanding this involvement, their primary research interests shifted somewhat to reflect the prevailing concerns of leading Nazi physicians and bureaucrats. Reche's political biases, evident in his previous work, were given full expression after 1933. In November of that year, Reche was among approximately 1,000 professors of German universities to pledge their commitment to Adolf Hitler and the National

---

Jews finally came to be a predominantly "Oriental mixed population," but with a strong "Armenian-Hamitic" makeup and lesser degrees of "Nordic, Mediterranean, Turanid, and Negro" influence. According to Wellisch, the Hebrews were already entirely "mixed" before the separation of the ancient empire into the southern Jewish state and the northern region occupied by the ten tribes. After their expulsion from Israel, they interacted with various peoples throughout the Diaspora. Initially, around the eleventh century AD, Western Europe (Moorish Spain, but also southern France and Germany), harbored most of the Jewish people. These communities continued to encounter groups from Asia and Africa. At the same time, the smaller eastern Diaspora was growing through influxes of Jews to the Byzantine Empire, Persia, and southern Russia and Poland. Jewish settlements in Russia, however, met Khazar and Cossack tribes. As a result, some of these Jews fled west, especially to Poland. This trend reversed in the thirteenth century and, over the course of the following two centuries, Jews were continually expelled from Western Europe. Ibid., 783–787.

[607] Ibid., 768.

Socialist state.[608] He also joined many National Socialist organizations—the *Nazionalsozialistiche Lehrerbunde* (National Socialist Teachers' Association), *Nazionalsozialistisches Kulturgemeinde* (National Socialist Cultural Community), *NSKOV* (National Socialist War Victim's Care), *Reichslufts-chutzbundes* (National Air Raid Protection League), *Rassenpolitisches Amtes* (Racial Policy Office), and the *Opferrings* (Circle of Martyrs).[609] Reche's racial activism would continue throughout the course of the Third Reich, and his career would benefit accordingly. In 1937 the Society for Physical Anthropology, of which Reche was a member, was reconstituted as the German Society for Racial Research.[610] Bruno K. Schultz and Wilhelm Giesler, both of whom had strong SS links, were appointed as directors, and Reche, along with Otmar Freiherr von Verschuer and Lothar Löffler, were on the society's committee.[611] In this role, Reche assisted the SS's organization for racial culture, the *Ahnenerbe*.[612] Whether a result of his principles, which were more quietly voiced during the Weimar Republic, or simple career opportunism—and most likely a combination thereof—Reche's anti-Semitism also grew more pronounced after 1933. He pursued more aggressively the matter of Siegmund Wellisch's racial identity. Wellisch was a member of the German Institute for Blood Group Research and a frequent contributor to its periodical—two of his articles had addressed the matter of "Jewish blood"—even though Reche strongly suspected that he was of Jewish descent. In the fall of 1935, Reche wrote to Steffan on the matter:

> Wellisch has sent me an excerpt from Eickstedt's *Zeitschrift für Rassenkunde*, in which he again addressed the Jewish problem—specifically, the total number of people with Jewish blood. He was always awkward...secretive [about his own race]. At the last conference in Berlin I met some men from Vienna, reliable National Socialists, who said that if someone from Vienna was named Siegmund Wellisch they would bet "a thousand to one" that he was certainly a pure Jew. This still needs to be researched. I am investigating the matter and hope to be able to inform you soon.[613]

---

[608] Geisenhainer, Rasse ist Schicksal, 179.
[609] Ibid.
[610] Weindling, *Health, Race and German Politics*, 503.
[611] Ibid.
[612] Ibid.
[613] Quoted in Geisenhainer, *Rasse ist Schicksal*, 198. "Egon Freiherr von Eickstedt developed a formula to establish people's race, supported by citations from Hitler and Rosenberg, with special attention to the correlation

Eventually, Reche did receive a response from Herbert Orel, a physician in Vienna. Orel explained that Wellisch's parents had married in the Jewish quarter of Vienna in 1890. Wellisch's father had converted to Christianity, and, though his mother remained Jewish, Wellisch and his siblings were raised Christian. Therefore, according to Reche, and more importantly the Nuremberg Laws, Wellisch was Jewish. However, Reche did not revoke his membership, though a brief article marking Reche's sixtieth birthday in 1939 credited him with having made efforts to remove "Jewish influence" from blood science through the German Institute for Blood Group Research and its periodical.[614]

The career of J.F. Lehmann, Reche's acquaintance, publisher, and co-editor of the *Zeitschrift für Rassenphysiologie*, also advanced during the Third Reich. In 1933 he was appointed to publish the first official commentaries to the 1933 Sterilization Law and the 1935 Nuremberg Laws.[615] He was rewarded for these contributions and in 1934 became the first member of the Nazi Party to receive its Golden Medal of Honor (*Goldene Ehrenzeichen*), a recognition bestowed only upon the original 100,000 members of the party.[616] Lehmann served as co-editor and publisher of the *Zeitschrift für Rassenphysiologie* from its beginning in 1928 until its final issue in 1943.

Considering the fact that both Lehmann and Reche edited the *Zeitschrift für Rassenphysiologie*, and the institute had a certain membership profile, one would have expected the majority of the articles published to have more closely reflected their own racial prejudices. Gauch's 1933 report on the relationship between blood type and race, with its references to "Nordic and non-Nordic" blood types, and their definitive correlations with physiognomy, seemed a better fit. Gauch further claimed that his subjects with type B blood did not have "any" Aryan characteristics. He expressed a certain confidence in seroanthropology that was lacking among his colleagues. Unlike Streng, Wellisch, and numerous others, Gauch did not hesitate in his claims or openly acknowledge the drawbacks of his research.

---

between race and character. His work *Rassenkunde und Rassengeschichte der Menschheit* (1934) ends with a call for eugenics, to fight the battle for the superior Nordic races against the backward southern stock. Benjamin H. Isaac, *The Invention of Racism in Classical Antiquity* (Princeton: Princeton University Press, 2006), 32, note 86.

[614] Geisenhainer, *Rasse ist Schicksal*, 133, note 87.

[615] Proctor, *Racial Hygiene*, 26.

[616] "Lehmann was granted this award even before such Nazi notables as Gerhard Wagner, Walther Darré, Justice-Minister Franz Gürtner, or Finance Minister Johannes Popitz." Proctor, *Racial Hygiene*, 27.

By the 1930s, such conviction was rare even in misappropriations of seroanthropology because, time after time, studies of blood and race had proven unreliable. Perhaps Gauch's research did not draw the attention of other German race theorists because his was not an "intellectual anti-Semitism," whereas Reche and Lehmann certainly believed that they were making respectable, professional contributions to German medicine through their racial research. By contrast, Gauch's work lacked similar scientific pretense. In a work entitled *New, Critical Issues in Racial Research*, Gauch claimed that "non-Nordic man" occupied an intermediate position "between Nordic Man and the animal kingdom, in particular the great apes...we could also call non-Nordic Man a Neanderthal: however, the term 'subhuman' is better and more appropriate."[617] The *Stürmer*-esque nature of Gauch's writings was not appreciated by more academic race theorists. Professor Löffler, a member of the Nazi Party and an open anti-Semite, as well as Reche's colleague in the German Institute for Blood Group Research, serves as a fitting example.[618] As chief of a race-policy bureau, Löffler prevented the distribution of Gauch's "bizarre and amateurish" 1933 book on racial research.[619] A concerted effort followed to prevent its distribution; instead of mere regional restrictions, Dr. Arthur Gütt, at the Reich Ministry of the Interior, requested that the book be nationally prohibited.[620] The Gestapo even became involved in confiscating existing copies of the work. One letter issued to a publisher in Leipzig explained that Gauch's work was "unsuitable" and felt it threatened public security.[621] As with his blood type research, Gauch's problem seemed to be with sweeping generalizations. Curiously, the Gestapo report explained that the German people had their origins not only in one race, as Gauch maintained, but were instead a "blend (*Gefüge*) of different races." Truly "racially pure" individuals, it concluded, hardly existed anymore.[622] Eventually, all of Gauch's publications were subject to censorship through the Racial Policy Office of the Nazi Party.[623] Restrictions were not enforced until the year after Gauch's

---

[617] Quoted in Müller-Hill, *Murderous Science*, 88.
[618] At one point, Löffler refused a chair in Frankfurt because of the city's large Jewish population. He subsequently accepted a chair in Königsberg, where he was appointed chief of the local race-policy bureau.
[619] Müller-Hill, *Murderous Science*, 88.
[620] R 58/904, 93. March 14, 1934, from the Reich Ministry of the Interior to the Prussian Ministry of the Interior.
[621] Ibid., 94. From the Gestapo to publisher Adolf Klein, March 16, 1934.
[622] Ibid.
[623] Ibid., 104.

article on blood and race was published in the *Zeitschrift für Rassenphysiologie*. Nonetheless, it was apparent even before these actions were taken that Gauch was not a respected race theorist within the German medical community, which raises the question of why the German Institute for Blood Group Research wanted to publish his work to begin with.

In fact, the *Zeitschrift* was often surprisingly objective. Instead of ignoring the harsh reality of seroanthropology's inconsistencies and shortcomings and censoring articles that described the ugly truth, the institute often served as an impartial reporter of current developments in blood science. Of those articles published after 1933, it was clear that blood science could not assist in determining an individual's racial type; it was only believed possible for blood to identify broad racial trends by blood type percentages that were completely inadequate for the Nazis' purposes of individual racial categorization. This was exacerbated by the fact that the authors themselves were often openly skeptical of seroanthropology and at times blatantly unsure of themselves. Even with the volume of research in place by the 1930s, Streng referred to seroanthropology as still being in its infancy and cautioned against making assumptions based on current statistics as, with time, these might turn out to be "completely erroneous."[624] He even theorized that it was possible for blood type to spontaneously change through mutation, although the legal admittance of the blood types in German courts certainly contradicted this.[625] To retain some semblance of respectability, the German Institute for Blood Group Research shifted more attention to non-racial applications of blood science.

It was clear that Reche did not want to lose Wellisch, whom he considered "brilliant" and whose work had been very valuable to the institute. As a Jew, Wellisch was what Reche referred to as their biggest "nut to crack." Perhaps this was not an obstacle that could not be overcome—Reche pointed out that other "foreign," or non-German, physicians had participated in medical conferences in the Third Reich. Obviously, he mused to Steffan, it was not possible to have a "racially foreign" individual serve in such a capacity as chair, but first he wanted to ask Dr. Walter Gross, the director of the Racial Policy Office, about allowing "official Jews."[626]

---

[624] Streng, "Die Bluteigenschaften (Blutgruppen) der Völker," 430.

[625] Oswald Streng, "Blutgruppenforschung und Anthropologie," *Zeitschrift für Rassenphysiologie* 9 (1937): 97.

[626] Ibid.

In spite of these discussions, Wellisch's racial status, and the political climate of the time, another article by Wellisch on blood and race was published in the 1937 edition of the *Zeitschrift für Rassenphysiologie*.[627] Likely on the advice of Reche and Steffan, Wellisch did not comment directly on the "Jewish Question." He believed that very few individuals were racially pure in their physiognomy; in addition, no "sharp line" could be drawn in racial classification.[628] Wellisch described the Nordic race as predisposed to "blond hair, light-colored eyes and skin, and considerable height"—but even in a Nordic group, he explained, these characteristics never reached a full 100 percent. An individual of brown hair and/or eyes could be classified as Aryan, provided that his or her remaining characteristics were "completely Nordic."[629] Wellisch reiterated his opinion that only combinations of different populations existed; there were no longer any "unmixed races." This applied to the Jews as well. In a separate article in the journal *Antropos*, Wellisch observed that efforts to trace Jewish descent through surveying the blood types of "Hebrews and Israelites" had been futile.[630]

During the interwar period, proponents of seroanthropology had repeatedly pointed to the unique blood type distributions of "native groups," such as Germans, Hungarians, or Roma-Sinti. This was one of seroanthropology's lingering attributes and its proponents would repeatedly refer back to it. However, an article published in the 1937 edition of the *Zeitschrift* questioned even this characteristic, as even the blood type profiles of native peoples could not be pinned down for comparison's sake. Estonian scientist Gerhard Rooks, for instance, reported significant variations in blood type patterns among "native Estonians." The first study, of 560 individuals from northern Estonia, indicated that they had 38.2 percent type O, 33.3 percent type A, 23.8 percent type B, and 4.7 percent type AB. Later, a separate exam-

---

[627] Wellisch, "Serologische Rassenanalyse," 38.

[628] Ibid., 37.

[629] Ibid.

[630] Ibid., 38. As for the Jews' "racial composition," their movements throughout the Diaspora indicated that they had come in touch with many different peoples. The Jews were tolerated longer in certain areas where they could establish "a foothold"; in Eastern Europe where the German-Polish Jews settled, and in Turkey, where the Spanish-Portuguese Jews settled. Wellisch claimed that the first group had mixed with Slavic, Tatar, Finnish-Ugric, Eastern European, Turanid, Alpine, and Mongolian peoples. As a result of their migration along the coast of North Africa and southern Europe, the Spanish-Portuguese Jews had been influenced by Mediterranean, Hamitic, Turanid, and Negro types. After centuries of isolation, Wellisch reiterated his prior claim that these two groups had developed into the Ashkenazi and Sephardi Jews and that the two were different from each other in both their physiognomies and blood type distributions. See Wellisch, "Rassendiagnose der Juden," 786.

iner supplemented these figures, increasing the number of Estonians examined to 849. This altered the resulting frequencies somewhat: O decreased to 34.2 percent, A remained the same, and both A and AB increased (to 26.2 percent and 6.3 percent, respectively). When another Estonian scientist, Aleksandr Paldrock tested an additional 200 leprous Estonians, the final percentages changed yet again. This time, O again decreased slightly, A jumped to 40.5 percent, B declined to 19 percent, and AB increased even further to 7.5 percent. It was clear from these findings, Rooks observed, that type distributions could vary from study to study.[631]

Oswald Streng also claimed that it was not possible to speak of constant blood type frequencies in one population or area. In one part of Germany, there had been a regular influx of mixed peoples. This was evident in the eastern parts of the nation along the Oder River, and was also the case between the Elbe and Oder, and in the southern regions of Bavaria, Baden, and Württemberg, but also in the Rhineland. Germany seemed to be a nation of "mixed peoples." The blood type statistics of western and southern Germans were found to clearly deviate from those in the east and central regions.[632] Blood type research frequently challenged the notion that the Germans were racially homogeneous, that the German people were a distinct racial type. Blood type percentages were so disparate between the eastern and western halves of Germany that Streng actually divided these Germans into two separate groups. This division, of course, contrasted with the Nazis' idyllic notion of German racial solidarity. Naturally, Streng's speculation that the Germans might have been of Asian descent similarly would have been anathema to Nordic racial theory.

Because of its inconsistencies and lack of precision, interest in seroanthropology steadily declined during the 1930s in Germany and elsewhere. (See Figure 17.) It would not have been practical for the Nazis to refer to seroanthropology in the construction and enforcement of the Nuremberg Laws, even though the legislation was based on disparities between individuals' blood. Those who had drafted the legislation, Stuckart and Lösener,

---

[631] Gerhard Rooks, "Über die Verteilung der Blutgruppen bei den Esten," *Zeitschrift für Rassenphysiologie* 9 (1937): 34. This inconsistency in findings, Rooks explained, probably stemmed in part from the fact that the Estonians examined were not all from the same region within the country. The blood scientist Vasilii Parin, who separately examined 140 Estonians living in Soviet Russia, found 30 percent had type O, 35.6 percent had type A, 29.3 percent type B, and 7.1 percent type AB.

[632] Streng, "Die Bluteigenschaften der Völker," 420.

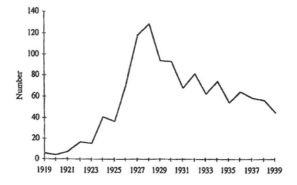

Figure 17. All (international) articles of original research on blood group distribution, 1919–1939. From William H. Schneider, "The History of Research on Blood Group Genetics: Initial Discovery and Diffusion," *History and Philosophy of the Life Sciences* 18, no. 3 (1996): 287.

had the unenviable task of converting malleable blood rhetoric into objective legislation. The Nazis' intent was to convey race as a scientific and precise category. However, they failed to do so, largely because of the fact that the murky association between blood and race lacked any scientific basis. When civil rights and lives hinged on whether one had "German blood," however, German bureaucrats had to explain themselves more thoroughly. This change had immediate ramifications for physicians, anthropologists, ministers, judges, and laypeople attempting to comply with the legislation. Ironically, the terms "Aryan" and "non-Aryan" had been deliberately replaced with "German-blooded" and "non-German-blooded" in order to prevent confusion. The result was that one ill-defined set of terms was replaced with another. Nonetheless, if the legal proceedings of the Third Reich serve as any indication, blood defilement through sexual or clinical means was believed to pose a very real threat. To prevent it, sexual contact between Germans and Jews was forbidden, as were blood transfusions—arguably the most tangible form of "blood mixing"; even here there was no way to differentiate between "German" and "Jewish blood." As the incident with Dr. Serelman had shown, "Jewish blood" could even be life-saving to an individual with "German blood."[633]

---

[633] 3NS2/152. "Personalbogen für Mitarbeiter." The form requests various information, and at the end, the following attestation: "I ensure that my wife and I are of Aryan descent and have no influence of foreign blood."

In spite of the Nazis' intent to protect German blood, and although a wide range of physical and mental characteristics was deemed acceptable for determining race, there is no indication that blood was considered useful in this capacity. Racial experts referred mainly to visible traits when deciding whether an individual had German blood. The regime's bias for physiognomy is obvious in an October 1935 letter issued from Dr. Bernhard Lösener to the Ministry of the Interior on the "solution of the half-Jewish question," written to advise racial experts on the classification of half-Jews:

> In the event that a half-Jew has exactly 50 percent Germanic and 50 percent Jewish blood, classifying them as either is sometimes not possible. In such a case, categorization results not from anthropological means but can only be accomplished through analysis of external appearance.[634]

Any individuals with a "strong Jewish appearance" were to be categorized accordingly, though Lösener conceded that even physiognomy was not foolproof, as every half-Jew possessed "thousands" of both Jewish and German characteristics.

Even though they preferred physiognomy, German race theorists, like Lösener, did on occasion also acknowledge its impracticality. Many Aryans had physical features the Nazis stereotyped as Jewish, and many Jews "looked Aryan." The Ministry of Propaganda even produced a film showing how easily Jews could physically pass as Aryans. In it "blue-eyed" Susanne, with the blond braids, who had just been praised in a racial theory class as a perfect example of the build and skull formation of the Nordic type, was abruptly expelled from school after being classified as 100 percent Jewish in accordance with the Nuremberg Laws.[635] In view of such exceptions, Lösener advised thorough consideration of other factors such as the individual's "disposition and psychology." Clearly, the final decision was based on personal opinion.[636] Even the state acknowledged that expert analyses could result in false racial classifications.[637]

The emphasis remained on physiognomy, even though it, like seroanthropology, was often unreliable and inconsistent. Studies of blood and

---

[634] R1501/5513, 141–142. To the Reich- and Prussian Ministry of the Interior, October 11, 1935.
[635] Miller, *Nazi Justiz*, 12.
[636] Schafft, *From Racism to Genocide*, 73.
[637] R1501/5513, 142.

race were similarly "flawed," yet were never advised by mainstream race theorists. The ambiguity of blood rhetoric still made it a very useful political tool, and loose propagandistic references to blood continued from the Weimar era until the end of the Third Reich. This continuity was evident in Nazi racial discourse from the vulgar anti-Semitism of lurid publications such as *Der Stürmer* to the highest levels of government. The Ministry of Propaganda explained:

> We know that blood is not simply a red fluid which pulses through our veins. Rather, it is our very being—which carries our ancestry and represents the lineage to which we will return. The same physical and spiritual predispositions are only found among men who are of the same blood. We are related to those who have carried the same blood. These carriers of the same blood—the different races—are different from one another.[638]

Similarly, the head of the Reich Chancellery explained that it was the "blood determined" inequality among humans in all aspects of their "thought, feeling and action" that resulted directly in their legal inequality.[639] Blood scientists, too, referred to blood in a metaphorical sense. In 1935 Otto Reche wrote:

> The blood of a gifted and culturally creative race flows in the veins of each individual German, even if they do not externally resemble the racial ideal; blood also contains the racial identity common to all Germans. The blood of the Nordic race is evident in our superior German culture.[640]

---

[638] NS2/162.

[639] Diemut Majer, *"Non-Germans" under the Third Reich: The Nazi Judicial and Administrative System in Germany and Occupied Eastern Europe* (Johns Hopkins University Press, 2003), 49–50.

[640] Quoted in Geisenhainer, *Rasse ist Schicksal*, 264.

# CHAPTER VII

## THE PEDAGOGY AND PRACTICE OF SEROANTHROPOLOGY DURING WORLD WAR II

Throughout the 1930s, the Third Reich focused primarily on stabilizing domestic matters in preparation for war. During this time, racial segregation in Germany continued to escalate; by 1939, more than 400 additional decrees, regulations, and amendments had consigned Jews and other "non-Aryan" groups to the outer fringes of society.[641] Under these circumstances, some Jewish blood scientists felt compelled to comment on their work, its larger political implications and misuse, and the role of race in ongoing research as Germany advanced towards war and the carrying out of the Final Solution. Nazi oppression relied mainly upon the biological notion of race to categorize individuals. It is important in this context to consider the state's opinion of seroanthropology, as the field was a branch of racial anthropology. Party-affiliated medical publications serve as one gauge of National Socialist interest in racial studies of blood. This literature indicates whether they were aware of past developments in seroanthropology, the nature of the differences the science revealed, and how it was related to conventional racial anthropology. The war had a definite impact on German medical objectives overall, and this was also evident in blood science. The war itself, occupation of foreign territories, and the cover provided by hostilities presented unique research opportunities that affected both racial and non-racial applications of serology.

---

[641] Anthony M. Platt, *Bloodlines: Recovering Hitler's Nuremberg Laws, from Patton's Trophy to Public Memorial* (Boulder, CO: Paradigm, 2005), 80.

In 1938 Ludwik Hirszfeld co-authored a book with his wife Hanna on blood science in biology, medicine, and law. In the book the Hirszfelds also commented on racial analyses of blood published since their 1918 study. The Hirszfelds were aware that societies and journals had formed in response, and they knew of the voluminous research that had followed to determine the serological makeup of a region's native peoples. By this point, Hirszfeld believed it "impossible" to cite all the studies.[642] He was aware of political misappropriations of seroanthropology and pointed to the Ukrainian Standing Commission on Blood Group Research as one example, as their objectives were dictated by the communist government. A state-appointed committee had heavily criticized the institute's previous research and instructed its directors, from that point forth, "to interpret all research according to the doctrines of Marx and Lenin."[643] Hirszfeld further likened this biased perspective to that of the German Institute for Blood Group Research, which he felt had been established to meet the needs of a "specific political demographic." This was especially obvious to him in a *Zeitschrift für Rassenphysiologie* article in which Hermann Gauch praised his colleague Paul Steffan for "having recognized the importance of serology for racial problems—particularly the study of the Jews, who have tried to 'pass' in silence and hide their racial identity."[644]

Hirszfeld refuted the institute's implicit notion of unique "Jewish blood." Instead, he argued, continuous movement and miscegenation throughout the Diaspora had resulted in diverse blood type distributions among Jews. Modern Jews were not racially homogeneous. In fact, Hirszfeld theorized that the Jews had already split into three (racial) types as early as 587 BC: Persian Jews (from the Babylonian exodus), the Jews of Kurdistan (descendants of the ten tribes of Israel), and the Jews of Yemen (who had arrived before the destruction of the First Temple). Their racial identity became even more complex as they migrated out of the Middle East and mixed with the local populations in Europe and throughout the Mediterranean Basin. Blood type surveys of Spanish, Polish, and

---

[642] Ludwik and Hanna Hirszfeld, *Les Groupes sanguins: Leur application à la biologie, à la médecine et au droit* (Paris: Masson, 1938), 130–131 and 138.

[643] Ibid., 151. This censoring occurred after the death of Dr. W.J. Rubaschkin in 1932, the founder of the institute.

[644] Ibid. This is in reference to Hermann Gauch, "Beitrag zum Zusammenhang zwischen Blutgruppe und Rasse," *Zeitschrift für Rassenphysiologie* 6 (1933): 116–117.

Russian Jews indicated that their blood type patterns differed by nationality. Hirszfeld argued that seroanthropology showed similarities between Jews and the majority population; for instance, the Jews living in Asia or Africa "more closely resembled the people of the Orient than the Jews in Europe."[645] To better determine why they so closely shared frequencies of blood types, Hirszfeld consulted his colleague Meir Balaban, a professor of Jewish studies at the University of Warsaw. Balaban believed he knew the source of these similarities:

> From the eleventh century, polygamy still existed among the Jews, and quite often slaves were taken as wives. In Constantinople and Rome, non-Jewish slaves were legally permitted to be sold into slavery. Because of this, prisoners of war were frequently circumcised and then sold as Jews. In addition, the Turks frequently attacked Poland and it was common for some (sometimes all) of the women in the villages, including the Jews, to be taken as prisoners and sold in the harems.[646]

The blood of the Jews was no different from that of surrounding peoples, yet the idea that it somehow was, was perpetuated by anti-Semitic researchers and National Socialists alike. Hirszfeld attributed misconceptions of literal blood differences between races to various social or political issues and criticized such misuse of scientific principles accordingly:

> I would like to separate myself from those who link the blood types to the mystique of race. We have created the notion of a serological race analogous to that of a biological race. A biological race is made up of a group of individuals who share a unique characteristic. The notion of a serological race *has nothing to do with that of an anthropological race.* Furthermore, the actual distribution of blood types across the globe indicates the mixing of races and provides even more proof that mankind is a mosaic of races. The anthropological races, by contrast, are characterized by a mix of arbitrarily chosen traits.[647]

---

[645] Hirszfeld and Hirszfeld, *Les Groupes sanguins*, 147.

[646] Ibid., 148–149. Balaban was a recognized historian of Polish Jewry and a prolific writer. Some of his works include *Jews of Lviv on the Eve of the Seventeenth Century* (1916), *Jewry of Lublin* (1919), *Jewish History and Culture with a Detailed Review of the History of the Jews in Poland* (1921), and *History of the Jews of Cracow* (1931). He was briefly a member of the *Judenrat* in the Warsaw Ghetto. Extract from the diary of Chaim A. Kaplan on the Warsaw *Judenrat*, 1941. Yad Vashem, Documents of the Holocaust. Balaban died in the ghetto in 1942. See Israel Gutman, *Resistance: The Warsaw Ghetto Uprising* (Boston: Houghton Mifflin Harcourt, 1998), 58.

[647] Hirszfeld and Hirszfeld, *Les Groupes sanguins*, 152.

This statement, however, seems to contradict previous comments made by Hirszfeld pertaining to blood and race. The metaphors of "pure" and "mixed" blood had been central to Hirszfeld's original 1918 study, in which he had grouped peoples by their "serological racial type."[648] Despite a heightened awareness of the danger of mixing politics and medicine, Hirszfeld's rhetoric seemed not to have changed. He reiterated how seroanthropological research had confirmed that the blood type distributions of people in Asia, Africa, and Europe were distinct from one another; a mere glance (*coup d'oeil*) at their findings made obvious the "enormous differences in the blood type frequencies of different peoples and races." [649] The four blood types, Hirszfeld reminded the reader, "always" (*toutefois*) corresponded to three serological racial types (types O, A, and B blood).[650] He also still claimed his biochemical index was precise in determining miscegenation, making it "easy" to distinguish the peoples of the Occident from those in Asia, and demonstrating, for instance, that the peoples of Denmark and Holland were "racially homogenous," while the Russians were not.[651] Only "racially pure" peoples would retain characteristic proportions of blood type, Hirszfeld explained, and this was most apparent in the serological matches between "the Germans in Russia and Germany, the English in Australia and England, the Dutch from the Transvaal and Holland, and blacks in America and Africa."[652] By contrast, mixed (*métisses*) groups were somewhere in between—miscegenation between "pure Indians" and whites had affected their descendants' racial identity just as it had the Germans of East Prussia and Saxony, whose elevated levels of type B blood indicated "past mixing with Slavs."[653] Hirszfeld referred to the German people as "unmixed," except for those in the far eastern provinces of Germany. He also continued to refer to type B blood as an "Eastern trait." In research conducted in his native Poland, Hirszfeld reported that type B blood was much more prominent in the south, a pattern he attributed to "pre-Indo-European influence and later Mongol invasions."[654] The tone of Hirszfeld's

---

[648] Myriam Spörri, "'Jüdisches Blut,'" 37–38.
[649] Hirszfeld and Hirszfeld, *Les Groupes sanguins*, 132 and 138.
[650] Ibid., 134.
[651] Ibid., 134 and 145.
[652] Ibid., 145.
[653] Ibid.
[654] Frank Heynick, *Jews and Medicine*, 438.

defense of seroanthropology seems incompatible with his criticism of political distortions of science, as both his analysis and terminology were so relevant to the concerns of *völkisch* scientists.

This perspective was also apparent in remarks made by German Jewish blood scientist Fritz Schiff who, in 1942, referred to Hirszfeld's work as "the greatest step forward since the introduction of the cranial index by Anders Retzius," and believed that one "could expect the study of blood groups to have, as it surely will in the future, a most prominent place in all anthropological investigations."[655] Schiff, who also studied seroanthropology, reiterated the fact that the highest frequencies of type B blood were in Central Asia, while type A blood was "fairly high everywhere," except for certain Native American tribes.[656] At the same time, Schiff denounced racism and claimed that any race distinguished on the basis of "some constellation of physical characteristics" was essentially artificial, as there were all sorts of "exceptions and gradations" among the racial types. According to Schiff, the fundamental units of racial variability were populations and genes, not "the set of characters which connote in the popular mind a racial distinction."[657] It is important to remember that both Hirszfeld and Schiff were intrigued by anthropological developments because medicine at this point was "saturated with biologically deterministic language and reductionist methodology," and it was a common assumption in all countries where researchers were examining human differences.[658] Both Hirszfeld and Schiff sought

---

[655]  Fritz Schiff and William C. Boyd, *Blood Grouping Technic: A Manual for Clinicians, Serologists, Anthropologists, and Students of Legal and Military Medicine* (New York: Interscience Publishers, 1942), 198. By far the most commonly used physiognomic measurement was the cephalic index—the relationship of the width of the skull to its length. It was first proposed by the Swedish anthropologist Anders Retzius in the 1840s as a way of distinguishing between the so-called *dolichocephalic* ("long-headed") blond Aryan peoples of Northern Europe and the supposedly inferior *brachycephalic* ("broad-headed") darker peoples of the south and east. See Schneider, *Quality and Quantity*, 217. There were only three reasons, Schiff believed, that explained why seroanthropology did not yet have this "prominent" role. First, studies of blood and race came relatively late: their geographic variation were discovered long after "other physical anthropological criteria" had enjoyed extensive use. Secondly, many anthropologists did not know enough about genetics to appreciate the advantages blood groups could offer for anthropological classification. The final reason, he theorized, was that the study of blood type frequencies often did not confirm existing notions about race. Schiff and Boyd, *Blood Grouping Technic*, 198.

[656]  Aside from what Schiff referred to as "several unconfirmed and random observations," the frequency of type B decreased as one moved away from these centers. By the time one reached America, Australia, and the more remote Pacific islands, the incidence of type B fell to zero. By contrast, Schiff explained, blood type A was "everywhere fairly high," except in certain Native American tribes, and seemed to reach its highest values in the "outlying corners of the world" ("refuge areas")—such as Scandinavia, Spain, and Australia. Ibid., 206.

[657]  Ibid., 202.

[658]  Efron, *Defenders of the Race*, 15.

to contribute to what was recognized as a legitimate science, and their discussion of blood's potential anthropological significance simply reflects the popular medical discourse of the time.

## SEROANTHROPOLOGY AND NATIONAL SOCIALIST MEDICINE

In a 1938 article featured in *Der Erbarzt* (The genetic doctor), Hans Weinert, a professor of anthropology at Kiel University, reported that the blood types could be linked to racially pure groups—their distribution among "modern Gypsies" was identical to that of Hindus in India. This phenomenon had also been reported in other peoples; just as all "Eskimos" had type O blood, so too did Native Americans—as long as they were, Weinert stipulated, also "racially pure."[659] Peter Dahr, director of the Institute for Serology at the University of Göttingen, made a similar observation in the publication *Ziel und Weg* (Road and way). Studies revealed that, even after several generations of separation from their homeland, the blood type distributions of ethnic Germans in Bessarabia were comparable to those of native Germans. Dahr maintained that *because they had not mixed*, the blood of the German settlers had remained completely different from the surrounding Russian people.[660]

Neither Weinert nor Dahr proposed entirely abandoning seroanthropology, however, and continued to cite the old, now very familiar justifications of *völkisch* blood scientists. Weinert was impressed by the fact that approximately 80 percent of the Nordic race had either type A or O blood. By contrast, he noted that type B occurred "only among some 10 percent," a statistic which led him to believe that type B had probably not existed among the *original* Nordic peoples. Weinert theorized, like Hirszfeld, that type B blood had been introduced into the Aryan population by Mongolians. This did not mean, he stressed, that type B indicated racial inferiority. Weinert was careful to point out that, despite its supposed Eastern origins, type B blood was also very common among the highest caste Indians—

---

[659] Hans Weinert, "Die anthropologische Bedeutung der Blutgruppen und das Problem ihrer Entstehung bei den Menschenrassen," *Der Erbarzt*, no. 10 (1938): 132.

[660] Dahr is referring to a study by M. Riethmüller in Bessarabia. See Peter Dahr, "Blutgruppenforschung und Rassenhygiene," *Ziel und Weg* 9 (1939): 106. See also Peter Dahr, "Neuere Ergebnisse und Probleme der Blutgruppenforschung," *Klinische Wochenschrift* 18, no. 35 (1939): 1173–1179.

those who had originally belonged to the European races and consequently were often "strongly Nordic." Because of this, he assured the reader, there was no need for a German with type B blood to feel "less worthy" (*minderwertig*). He further remarked that correlations between blood types and behavioral traits were completely unfounded.[661]

Printed in the popular National Socialist medical periodicals *Der Erbarzt* and *Ziel und Weg*, Weinert and Dahr's assessments of seroanthropology would have been distributed throughout Germany. *Der Erbarzt*, designed specifically to address *völkisch* racial concerns, first appeared in the summer of 1934.[662] The editor, devoted Nazi physician Otmar Freiherr von Verschuer, described *Der Erbarzt* as a response to the "revolution of 1933," and a link between "the ministries of public health, the genetic health courts, and the German medical community."[663] In simpler terms, however, its main purpose was to provide a forum for discussion of methods, criteria, and grounds for imposition of negative eugenic measures, such as sterilization.[664] Though its overall content was more diverse, *Ziel und Weg* was not much different. Established in 1931 as the official journal of the National Socialist Physician's League, it served as the medical outlet of the Nazi Party.[665] Both publications were created to address the clinical, and thereby largely eugenic and racial concerns of the National Socialist Party. Weinert and Dahr's articles indicate that the scientific research of blood and race had not gone entirely unnoticed by the Nazi medical community. However, each author ultimately denied seroanthropology's usefulness.

Both Weinert and Dahr had been careful to strongly emphasize that the relationship between blood and race was tenuous. Weinert acknowledged that the proportions of the blood types could vary between different groups, but this did not necessarily signify racial variation, as blood type distributions for widely divergent lands and peoples had, on occasion, been very close, if not the same. Similarities between the blood of peoples traditionally viewed as entirely separate racial types—like Afri-

---

[661] Ibid. Because of its infrequency (occurring in usually less than 5 percent of the population), the remaining type AB blood was often not even mentioned.

[662] Proctor, *Racial Hygiene*, 104.

[663] Ibid.

[664] Ibid.

[665] Robert J. Lifton, *The Nazi Doctors: Medical Killing and the Psychology of Genocide* (New York: Basic Books, 1986), 32.

cans and Europeans—were very damaging to the perception of blood type as a reliable racial indicator. Furthermore, for the time being, serological type was basically irrelevant, as the race of an individual could not be determined from his or her blood. Nor could the blood of a Jew be distinguished from that of a non-Jew. Nor, for that matter, could one type definitively labeled as "primitive" or indicative of an "inferior race." As Dahr summarized it, "when the members of the Nordic race have a predominance of blood type A coupled with a significant decline in type B, it cannot be said that only type A is found within the Nordic race."[666] The coup de grâce was certainly the fact that it was not possible to identify an individual's race through his or her blood. Despite the "best expectations and efforts," Weinert complained, blood typing had failed to produce proof of "Aryan descent."[667]

The unknown origins of the blood types were yet another drawback. Without these, Weinert explained that it would be "extremely difficult" to make any correlations whatsoever between blood type and race.[668] He had first brought this matter to readers' attention years earlier in an article in the *Zeitschrift für Rassenphysiologie* in which Weinert had raised the question of whether the problem of the "missing link" might be clarified by crossing humans and apes, since "the evidence of living bodies [has] more to say than the best fossils of extinct transitional forms." At that point, he even proposed inseminating a female chimpanzee with the sperm of an "African Negro, preferably a jungle pygmy."[669] He speculated that the "inequal-

---

[666] "Lebenspende," *Neues Volk* 8 (August 1940): 27–28.

[667] Weinert, "Die anthropologische Bedeutung der Blutgruppen und das Problem ihrer Entstehung bei den Menschenrassen," 132.

[668] To research evolution, some researchers had tried to determine the blood type distributions of primates. After years of such research, however, it seemed that this approach, too, failed to provide the anticipated insight lacking in corresponding studies of humans. Weinert was well aware of these inconclusive results—even after a "thorough overview" of the blood type distributions of anthropoids and humans, the question as to whether mankind originally shared one common blood group still remained. Because this was not known, Weinert further explained that the blood types were not capable of addressing the matter of polygenism (racial diversity) in human beings. In the early twentieth century, it was commonly believed that Asians had descended from orangutans, blacks from gorillas, and Europeans from chimpanzees. This notion, Weinert declared, had been "completely refuted" by the fact that the distribution of the four blood types was comparable to that within the human population. The four serological types were still found in each primate group studied and, as had been the case with human subject groups, there were inexplicable patterns in their distribution frequencies (i.e., comparable results between different primates, just as there were between Africans and Europeans). Ibid., 133.

[669] "The Soviet biologist Ilja I. Iwanow, a specialist for the artificial insemination in animal breeding and for experiments on hybridization between species, had tested this very idea in practice in 1926/27 on an expedition to West Africa financed by the Soviet government. Iwanow had even conducted negotiations with the

ity" of the human races could be attributed, in part, to the "fact" that the various peoples of the earth had evolved at different rates. Included among the "less-evolved" types were "African Negroes" and Australian Aborigines. As the first to evolve, Weinert explained, Africans had given rise to other humans, but this also meant that they had remained "retarded."[670] Working from the premise that the origins of the different racial types were critical to modern racial classification, he presented what would have been a compelling case against using blood for racial analysis. As it was not known where, when, or how the various blood types had evolved, Weinert argued that they were basically useless and said as much in *Der Erbarzt*.

In *Der Erbarzt* Weinert touched on a recent discussion in studies of blood and race concerning which of the blood types was the "most primitive." Type A was commonly affiliated with "Nordic" descent. But, increasingly, so was type O, even though it was the people of Sardinia, not Northern Europeans, who had the highest frequency of this type in Europe—nearly half of the population.[671] Whereas both Weinert and Gauch referred to type O as "Aryan," others came to associate it with those who were "racially inferior." In an article published by the German Institute for Blood Group Research, examiner Joachim Richter proposed that type O individuals were predisposed to tubercular infection because of their type's origin in "primitive and isolated peoples."[672] In France, Nicholas Kossovitch also theorized that type O blood might indicate a "primitive constitution," as it was very common among such groups as the "American Indians, Eskimos, and Malay of the Philippines."[673] Similarly, Swedish blood scientist Otto Streng claimed that type O was the first group, based

---

governor of Guinea and the physician of the hospital in Konakry in order to procure permission to artificially inseminate African women with the sperm of chimpanzees." Hans Walter Schmuhl, *The Kaiser Wilhelm Institute*, 83 and footnote 205. See also Hans Weinert, "Blutgruppenforschung an Menschenaffen und ihre stammesgeschtichtliche Bewertung," *Zeitschrift für Rassenphysiologie* 4 (1931): 8–23.

[670] Robert Proctor, "From Anthropologie to Rassenkunde in the German Anthropological Tradition," in George W. Stocking, ed., *Bones, Bodies, and Behavior: Essays in Behavioral Anthropology* (Madison, WI: University of Wisconsin Press, 1990), 152.

[671] Ibid.

[672] Joachim Richter, "Die Blutgruppen- und Blutfaktorenverteilung in Nordhannover," *Zeitschrift für Rassenphysiologie* 11 (1939): 16. Joachim Richter suggested that individuals with type O were especially susceptible to tuberculosis, as an English study had found an elevated frequency of this type among tubercular subjects examined.

[673] Raymond Dujarric de la Riviere and Kossovitch, "Les groupes sanguins en Anthropologie," 277.

on its high incidence among the most "primitive peoples" of the globe.[674] German physician Hans Sachs similarly entertained the idea that type O came first, with A and B being the result of "later mutations." As evidence, he referred to living peoples and the results of blood type examinations of Native American mummies.[675] Hirszfeld interpreted the low frequencies of types A and B blood among Native Americans as "proof" (preuve) that blood type O represented "the most primitive human races."[676] Based on paleontological and blood type evidence in South Africa, British anthropologist Reginald R. Gates believed that the ancestors of the Bushmen of South Africa had type A blood, which would have made it older and therefore "less evolved" than type B blood.[677] American physicians Wyman and Boyd claimed that type A individuals had been driven out to refuge areas by the "superior strength, skill, or numbers of [those with] type B."[678] Based on high percentages of type A blood among Aborigines, another physician argued against the common theory that type A originated in Eurasia; similar statistics had been reported in Polynesia and New Zealand, as well as other islands in the Pacific.[679] This lack of consensus concerning the evolution of the blood types, mentioned by Weinert, was yet another source of frustration in seroanthropology that would have discouraged further Nazi pursuit of seroanthropology.

A 1940 article in *Neues Volk* (New Nation), a monthly publication of the Third Reich's Racial Policy Office and an important mouthpiece of Nazi medicine, also effectively disregarded seroanthropology—despite a reference to the "interesting" finding that blood type frequencies varied among different peoples, and the "striking" pattern in which blood type A became less common as one moved eastwards. These differences were most pronounced, the article explained, if one compared the blood type distributions in Germany to those in India. In the Rhineland, only about 8 percent of the population had type B blood, whereas the frequency was much

---

[674] Streng, "Die Bluteigenschaften (Blutgruppen) der Völker," 424.

[675] Hans Sachs, "Blutgruppen, Bluttransfusion und Abstammung," *Practica oto-rhino-laryngologica* (1939): 39. Matson's research revealed type O in sixteen Native American mummies, whereas only three of six Egyptian mummies belonged to type O.

[676] Hirszfeld and Hirszfeld, *Les Groupes Sanguins*, 149.

[677] R.R. Gates, "Recent Progress in Blood Group Investigation," *Genetica* 18 (1935): 52–53.

[678] If its mutation frequency were sufficiently high or if it were favored (over O) either directly or indirectly by natural selection. Ibid., 53.

[679] Ibid., 428.

higher in India.[680] However, this article, too, made reference to the "discrepancies" in blood type research—specifically, elevated frequencies of the purportedly Aryan type A blood that had been reported in Australia and the South Seas. In addition, high percentages of type B had been found among peoples on the southernmost tip of South America.[681] Seroanthropology was both "interesting and exciting" but also inexplicable and therefore useless for purposes of racial differentiation. Instead, the pages of *Neues Volk* were littered with illustrations depicting "racially pure Aryans" opposite "inferior" types. Appearance was of utmost importance, while to the authors "blood" remained simply a metaphor:

> The life-giving bodily fluid of blood has a symbolic meaning outside of the body. We say that particularly capable or leading men are of noble blood, or we speak of "blood and soil" when referring to the fateful relationship between race and living space. Finally, blood has in another sense the meaning of race or heredity, in which it is an actual carrier of hereditary disposition. Today, the little word blood is most often used in this sense.[682]

*Neues Volk* had an even wider circulation than *Der Erbarzt* and *Ziel und Weg*; in one calendar year, 750,000 copies were distributed.[683] It was delivered to all doctors, dentists, and pharmacists as reading material for waiting rooms.[684] It could also be found in schools, public libraries, and private homes.[685]

## SEROANTHROPOLOGICAL RESEARCH

While conventional surveys of blood type continued to decline, the German Institute for Blood Group Research did employ this method to revisit a theme popular during the Weimar Republic and within National Socialism—the notion that German peasants were the most "racially pure" representatives of the "Nordic type." In September 1933, the Hereditary Farm

---

[680] Frequencies listed for the Rhineland: type A 44 percent, type B 8 percent, type AB 3 percent, type O 45 percent.
[681] "Lebenspende," *Neues Volk* (1940), 27–29.
[682] Ibid., 26.
[683] Roger Uhle, "Neues Volk und reine Rasse. Walter Gross und das Rassenpolitische Amt der NSDAP (RPA)," dissertation (Technische Hochschule Aachen, 1999; 47.
[684] Carol Poore, *Disability in Twentieth-Century German Culture* (Ann Arbor, MI: University of Michigan Press, 2007), 102.
[685] Ibid.

Law was passed to implement this ideology. It excluded Jews from owning farmland or engaging in agriculture.[686] Though this had little effect upon the Jewish community, Reich officials explained that the law was necessary to "secure the peasant foundations of our blood through instituting the ancient customs of land inheritance."[687] Furthermore, the German farmer had to be able to trace his racial purity back to the year 1800, before the emancipation of West European Jewry.

During the 1920s, many studies sought to identify native or peasant Germans in such isolated locations as remote farmland, forests, and even islands in the North Sea. Researchers also consulted historical documents to make certain that their subjects had not been exposed to "non-German blood." At that time, the results were not consistent enough to elicit much interest. Nonetheless, nearly identical analyses were repeated in the late 1930s. A 1939 *Zeitschrift* article reported on the blood type distributions of Germans outside of Hanover and Hamburg, areas chosen—like others before them—for their "isolation from traffic."[688] The researcher, Joachim Richter, explained that miscegenation was unlikely, given the area's geography and history: Hanover was bordered by the Weser and Elbe rivers, and Hamburg was surrounded by floodplains. Richter anticipated that the racial purity of the subject group, an "indigenous group of farmers who had tilled the land for generations," would be apparent in their blood.[689] However, the results were inconclusive, with no significantly higher frequency of one blood type. Nor was there a pattern between serological type and other characteristics analyzed, such as syphilis.[690]

Researcher Paul Schmidt also selected a group in Schleswig-Holstein, as this was "the part of the Fatherland (*Vaterland*) with the strongest Nordic influence."[691] Unexpectedly high levels of Eastern blood type B were found, however, which contradicted this claim. Schmidt attributed this result to miscegenation in the twelfth century between the Germans and "Wends,"

---

[686] Schleuenes, 112.

[687] Schleunes, *The Twisted Road to Auschwitz*, 112.

[688] Richter, "Die Blutgruppen- und Blutfaktorenverteilung in Nordhannover," 15–21.

[689] Ibid.

[690] Ibid., 16. Despite inconclusive results, Richter still believed that the "exercise" had been helpful, as it helped organize the regional types for blood donation purposes.

[691] Paul Schmidt, "Die relative Haüfigkeit der Blutfaktoren M und N und der 'Untergruppen' A$_1$ und A$_2$ unter besonderer Berücksichtigung Schleswig-Holsteins," *Zeitschrift für Rassenphysiologie* 11 (1940): 49.

a Slavic minority scattered throughout northern and eastern medieval Germany.[692] As similar Weimar-era studies had done, he acknowledged that past mixing was likely, but still declared that his subjects were still "predominantly Nordic." A near-identical interpretation was presented by Schlossberger et al. in a 1926 study of native Germans near Frankfurt am Main.[693] Furthermore, all of these studies, both old and new, were focused on what were, in effect, disappointingly minor differences. For instance, Schmidt sought to determine why one village in Schleswig-Holstein had a lower incidence of type A blood than another—one had 45.5 percent, the other 48.2 percent. This was a familiar drawback of seroanthropology, a science that required exaggeration in order to be used as evidence of racial difference.

An overview of the collected German blood showed significant incon-sistencies in results. Despite the fact that both Schmidt and Kerkhoff's articles were included in the *Zeitschrift für Rassenphysiologie*, Hans Kerk-hoff's figures further contradicted the notion that Schleswig-Holstein was "predominantly Nordic" in its elevated levels of type A blood. In fact, he observed a relatively high percentage of type B blood in northern Ger-many—in some areas, it was 20 percent. In addition, Kerkhoff reported that type A was at its highest in western Germany at 45.92 percent, though Schmidt claimed that North Friesland had 48.2 percent. Furthermore, while the national average of type A blood was 43.62 percent, the high-est frequencies of type A blood were found outside Germany.[694] At 52.4 percent, the Swedish Lapps had the most type A of any people tested.[695] Because type A was on average only slightly higher than 40 percent among the German people, this meant that the majority of the German people had a different (non-Nordic) blood type. Frustrated by the various percentages

---

[692] Ibid., 53.

[693] Ibid. From his own figures, Schmidt reported that both the Schleswig-Holstein areas, Dithmarschen and Hol-stenland-Stormann, still had high percentages of type A (45.5 and 45.7 percent, respectively). However, these were not as high as some other figures reported in the area. Dithmarschen, he theorized, had probably re-ceived a stronger influx of "Frisian blood," while Holsten-Stormann had been settled by Saxon Holsten and Stormann tribal groups. According to Schmidt, these overall "lower levels" of type A (relative to others near-by) was the result of "centuries of immigration" and also, in the case of Holstenland-Stormann, "heavy indus-trialization." Ibid.

[694] Germany's averages: type O 38.26 percent, type A 43.62 percent, type B 13.04 percent, type AB 5.08 percent. Kerkhoff, "Blutgruppenuntersuchungen in einem Eifeldorf: ein Beitrag zur Frage der Vererbung der Blutei-genschaften ABO und MN und zu ihrer Verteilung in der Bevölkerung," 33.

[695] Ibid., 38.

throughout, some even divided Germany into two or four zones according to their disparate blood type frequencies.[696] Even this was ultimately not helpful, however, as there were blurry transition areas between these zones, so no sharp division between the types really existed.[697]

Despite the sheer volume of blood type surveys during the interwar period—in 1941 one author reported that 56,062 blood samples of the traditional ABO types had been collected in western Germany alone—the indefinite nature of this technique prompted efforts to move beyond it.[698] Peter Dahr believed that scientists should more closely examine the serological characteristics of M, N, $N_1$ $A_1$, and $A_2$ traits, which were similar to blood type and the MNP characteristics discovered by Landsteiner and Levine in New York—these, too, were hereditary, protein-based differences in the blood.[699] Since the 1920s, researchers had begun to examine their frequencies in large groups of peoples, though not on the scale of the mass blood type surveys.[700] Several articles in the 1940 edition of the *Zeitschrift für Rassenphysiologie* were on the heredity and frequency of these traits.[701] Even preliminary results, however, suggested that this methodology was fraught with the same inconsistencies and contradictions of conventional blood type surveys.[702] Other researchers, frustrated by the imprecision of the survey technique, focused their efforts instead on determining race from an individual sample of blood.

During the war, German anthropologist Karl Horneck explored the possible existence of what he referred to as "race-specific proteins" in the

---

[696] Kerkhoff divided Germany into four serological zones.

[697] A.S. Wiener, *Blood Groups and Transfusion* (Waukesha, WI: Thomas, 1943), 295.

[698] Hans Kerkhoff, "Blutgruppenuntersuchungen in einem Eifeldorf: ein Beitrag zur Frage der Vererbung der Bluteigenschaften ABO und MN und zu ihrer Verteilung in der Bevölkerung," *Zeitschrift für Rassenphysiologie* 12 (1941): 32. Of these, 41.37 percent had type O blood, 45.9 percent type A, 9.29 percent type B, and 3.41 percent type AB.

[699] Peter Dahr, "Blutgruppenforschung und Rassenhygiene." *Ziel und Weg* 9 (1939): 104-105.

[700] Schockaert found N in his Belgian subjects, Amzel M in the Polish, Schigeno both M and N in the Japanese, and Thomsen noted the same among the Danish. In 1931, Wiemer and Vaisberg carried out research on the heredity of M and N. See Steffan, ed., *Handbuch*, 15.

[701] See V. Friedenreich, "Einige Bemerkungen zur Frage der Vererbungsweise der $A_1$ $A_2$ Eigenschaften," *Zeitschrift für Rassenphysiologie* 11 (1940): 22–24; Paul Schmidt. "Die relative Haüfigkeit der Blutfaktoren M und N und der 'Untergruppen' $A_1$ and $A_2$ unter besonderer Berücksichtigung Schleswig-Holsteins." *Zeitschrift für Rassenphysiologie* 11 (1940): 49–77 ; P. Dahr, H. Offe, and H. Weber, "Weitere Erblichkeitsuntersuchungen über den Blutfaktor P bei Familien und Zwillingen," *Zeitschrift für Rassenphysiologie* 11 (1940): 78–92; Gerhard Rooks, "Über die Verteilung der Blutgruppen bei den Esten," *Zeitschrift für Rassenphysiologie* 9 (1937): 33.

[702] Between 1929 and 1930, physician Fritz Schiff examined the M and N traits of Berliners and observed that they were "similar to Americans."

blood. With the cooperation and support of the staff of the General Hospital of Le Havre, in German-occupied France, Horneck arranged to have blood samples drawn from prisoners of war detained in the area. Most of the men were colonial subjects of non-European descent who had been brought to France at the onset of hostilities. From each sample, Horneck isolated the sera, a blood derivative, which he then injected into rabbits to observe if the animals would react differently to "European, Moroccan, Indochinese, and Negro" blood.[703] Horneck described the results as "striking and unexpected"; the sera of the white men prompted a faster and stronger reaction, which led Horneck to theorize that European sera contained less of a certain protein than the other races tested.[704] He was confident that he had made critical progress towards the possibility of racial identification and, based on these findings, claimed that it was now necessary to analyze these proteins when making "a blood-based racial diagnosis"—even though there was no such blood-based method of racial classification for it to replace.[705]

The possibility that unique racial proteins existed in the blood piqued the interest of Berlin anthropologist von Verschuer, who was well aware of the limitations of conventional seroanthropology and voiced the usual criticisms. Though they had been "all the rage for a while," racial studies of blood had not presented any new or improved means of racial differentiation; only the proportion of blood types in groups was "somewhat relevant." Von Verschuer noted a somewhat higher proportion of blood type B among Jews, which he believed to be "in keeping with our concept of the racial origins of the Jews as somewhere between the Near Eastern and Oriental groups."[706] Nonetheless, it was "still not possible to distinguish between Jews and non-Jews from their blood." Von Verschuer referred to the efforts of E.O. Manoiloff to do so in the 1920s, though the fact that nothing had been published on that topic in nearly twenty years confirmed to him that this method had not been effective.[707] Von Verschuer was a

---

[703] Horneck specifically thanked Professor Vincent, his lab assistant Frau Dautry, and Dr. Boehm for making the study possible. Karl G. Horneck, "Über den Nachweis serologischer Verschiedenheiten der menschlichen Rasse," *Zeitschrift für menschliche Vererbungs- und Konstruktionslehre* 26, no. 3 (1942): 309.

[704] Ibid., 311 and 316.

[705] It was necessary to determine both the protein content and the "albumin-globulin quotient." Ibid., 318.

[706] Otmar Freiherr von Verschuer, "Rassenbiologie der Juden?" *Forschungen zur Judenfrage* 3 (1938): 137–151.

[707] Ibid., 142.

highly respected and influential medical figure during the Third Reich. Importantly, he considered seroanthropology in its then-current state to have very little potential, though he was willing to pursue further analysis of race and blood through a technique besides surveying the blood types or Manoiloff's blood-type technique.

In August 1943, von Verschuer applied for and received a grant from the German Research Council (*Deutsche Forschungsgemeinschaft*) to study the existence of racial proteins in individuals suffering from infectious disease.[708] The project was approved by *Reichsführer-SS* and Chief of German Police Heinrich Himmler.[709] In his progress report of March 20, 1944, von Verschuer explained that he had been able to collect racially diverse blood samples with the assistance of Josef Mengele, a camp physician and SS captain at Auschwitz.[710] In the camp, Mengele had selected identical and fraternal Jewish and "Gypsy" twins with the same quantity of typhoid bacteria, drawn blood at various times, and followed the course of the disease.[711] Survivors later reflected on the surprising amount of blood taken: "We were wondering where [the blood] came from"; and toward the end they remembered it as being difficult to draw—"it wasn't coming anymore… from our arms."[712] The samples were shipped back periodically to von Verschuer at the Kaiser Wilhelm Institute of Anthropology, Human Heredity, and Eugenics in Berlin. They were specially packed and stamped "War Material—Urgent," such parcels were given top priority in transit.[713] In October of that same year, von Verschuer reported to a colleague that more than 200 samples had been gathered from prisoners at Auschwitz "of various races, some twin pairs, and some families," which meant that the "real research" could begin very soon.[714]

---

[708] Lifton, *The Nazi Doctors*, 341.

[709] Müller-Hill, *Murderous Science*, 78.

[710] Mengele was transferred to Auschwitz as a camp doctor on May 30, 1943. Mengele's first duty was to control the epidemic of typhoid brought in by the Gypsies who had been transferred from Bialystok in March of that year. He "selected" several hundred of the sick and sent them to the gas chambers. Ibid., 76.

[711] Ibid., 78.

[712] Lifton, *The Nazi Doctors*, 350.

[713] Müller-Hill, *Murderous Science*, 99. The Nazis took advantage of the racial diversity in the camps. With the approval of Himmler, many anthropological studies were carried out. In addition to the "specific proteins" project, the Reich Research Council also approved another on "eye color"; Mengele sent both blood and eye samples back to von Verschuer for analysis. Ibid., 20.

[714] Ibid., 79.

Bruno Weber, chief of the Hygienic Institute at Auschwitz, also carried out experiments involving blood on camp inmates. Weber typed the blood of some and injected others with incompatible types in order to study the harmful effects of agglutination, or clumping, of blood cells.[715] Mismatched blood types cause red blood cells to rupture in the recipient; the painful symptoms of receiving a non-compatible type can take many forms, including fever, chills, chest pain, and even death.[716] Physicians at Auschwitz collected blood in sometimes-lethal amounts, whether for experimental use or, apparently, for use in transfusions for German personnel—despite the "non-German blood" of the donor. After the war, one camp doctor claimed that he and his colleagues had taken (small amounts of) blood for the practical purpose of producing test sera, which was needed in order to type blood for transfusions. However, he also admitted that Weber had instructed SS men to "go to the camp, fetch yourselves a few fat *capos*, and tap [*zapft*] them," presumably to acquire much-needed blood for the Eastern Front.[717] SS men did not limit their selection to well-nourished inmates but "tapped blood wherever they could get hold of it because it was much less work that way."[718] In late 1944 the Germans were forced to abandon their experiments and forcible drawing of blood as the Soviets neared Auschwitz.

A 1943 article published in the *Zeitschrift für Rassenphysiologie* indicates that other German researchers also exploited the circumstances of war to conduct seroanthropological research. This study of men detained in a German prisoner-of-war camp in North Africa similarly diverged from the usual survey-based approach. Approximately 18,000 prisoners were held in the camp, from countries throughout Europe, Africa, Asia, and even South America, though the researchers only examined a fraction of this number—certainly due to time constraints.[719] Hoping to discern some pattern

---

[715] Lifton, *The Nazi Doctors*, 289.

[716] Jeffrey A. Norton, *Surgery: Basic Science and Clinical Evidence* (New York: Springer, 2001), 177.

[717] Lifton, *The Nazi Doctors*, 289.

[718] Ibid., 289.

[719] Blood samples, the authors declared, had been collected only from healthy individuals, who were reportedly not hard to find within the "spacious and orderly" accommodations of the camp, which "complied with all hygienic requirements." French served as the common language. The North Africans were even fluent, the blacks partially so; for all others an interpreter was used. See Robert Stigler et al., "Rassenphysiologische Untersuchungen an farbigen Kriegsgefangenen in einem Kriegsgefangenenlager," *Zeitschrift für Rassenphysiologie* 13 (1943): 26–27.

among the "racial types" represented, the physicians carefully recorded how long it took a subject's blood to coagulate after being drawn. Repeated tests of the same fifty black subjects revealed that their blood, which the Germans referred to as "especially thick and sticky," also clumped "surprisingly faster" than the white samples.[720] Under closer scrutiny, however, the description "surprisingly faster" becomes much less impressive; on average, the blood of the blacks coagulated in two to nine minutes, whereas that of the whites ranged from six to fourteen minutes. Analysis of other traits proved similarly disappointing. The blood of the Europeans was, on average, "warmer than the blacks by about two degrees." But this was not consistent, or easily tested, as it was often necessary to "stick the Africans much deeper" for a sample of blood—one examiner remarked on how their blood vessels did not seem to be as close to the surface. After "careful analysis," the physicians decided that none of these traits could be used in assessing racial difference.[721] The physicians abandoned their experiments with blood and moved on to other potential indicators of race, including "reaction time, frequency of left-handedness, patterns of body hair, and even size of genitals."

## SEROANTHROPOLOGY AND NAZI RACIAL IDEOLOGY

Even though blood science was not used during the Third Reich to determine the presence of "German blood," blood rhetoric, especially the recurring theme of blood defilement, continued to play an important role in Nazi racial discourse. In *Mein Kampf*, Hitler had written that the main purpose of academics was to "burn the racial sense and racial feeling into the instinct and the intellect, the heart and brain of youth entrusted to it. No boy or girl must leave school without having been led to an ultimate realization of the necessity and essence of blood purity.[722]

In Nazi Germany, texts written for instruction in schools were modified according to Hitler's worldview. Alfred Vogel, a biology curriculum writer

---

[720] Stigler examined the coagulation rate of the blood of "thirty Arabs, thirty-nine Indochinese, four American and eight African Negroes." The blood temperature of the Europeans was on average 13.95 degrees. The average temperature of the African Negroes was, however, 11.79 degrees. Ibid., 36. The fifty blacks were thirty Tonkinese, ten Senegalese, and ten Moroccans. Ibid., 27.

[721] Ibid., 33.

[722] Adolf Hitler, *Mein Kampf*, 427.

and elementary school principal, and Karl Bareth, a teacher, co-authored *Heredity and Racial Science for Elementary and Secondary Schools* (*Erblehre und Rassenkunde für die Grund—und Hauptschule*). Published in 1937, the text provided elementary-school teachers with instruction on explaining genetics and race. The table of contents reveals that, even at this young age, blood rhetoric was applied in an academic setting, with sections on "The Nordic race as the basis of blood for German people," and "Keeping the blood pure."[723] Another Nazi primer, given to German boys and girls between the ages of fourteen and eighteen, claimed that possession of "German blood" was essential for acceptance into the community of German people.[724] These were only metaphors, of course, and no specific reference was made to serology. When determining race, Nazi race theory consistently emphasized not blood but physiognomy. Various traits, such as the supposedly more elongated skull of the German people, were cited as visible proof of their "superiority" over other races. However, most race theorists did not refer to only one trait, even the "critical" shape of the skull, in racial categorization. Vogel's attention to the "elongated German skull" was offset by a separate six pages explaining distinct skull shapes among the numerous German races.[725]

The importance of protecting the "purity" of one's blood was repeatedly impressed upon students; a race would retain its purity and characteristics only if it did not mix with others. Nazis believed race to be a genetic phenomenon, passed down from parents to children and not subject to

[723] Alfred Vogel, *Erblehre und Rassenkunde für die Grund- und Hauptschule* (Baden: Konkordia, 1937), 74 and 96. See also Gregory P. Wegner, "Schooling for a New Mythos: Race, Anti-Semitism and the Curriculum Materials of a Nazism Race Educator," *Paedagogica Historica: International Journal of the History of Education* 27 (1991): 189–213.

[724] Fritz Brennecke, *The Nazi Primer: Official Handbook for Schooling the Hitler Youth* (New York: Harper and Bros. Publishers, 1938) and Fritz Brennecke, ed., *Handbuch für die Schulungsarbeit in der HJ. Vom deutschen Volk und seinem Lebensraum* (Munich: Zentralverlag der NSDAP, 1937), 13.

[725] Gregory Paul Wegner, *Anti-Semitism and Schooling under the Third Reich* (London: Routledge Falmer, 2002), 76. Anthropologists commonly claimed that the German people were made up of various "racial types." Albrecht Wirth described the German *Volk* as originally made up of "Celtic, Lithuanian, Germanic and Slavic blood, then Romans, Jews, Huguenots and Italians, with a smattering of Swedes, Scots, Croats, Irish, Hungarians, Spaniards and Turks." As quoted in Hutton, *Race and the Third Reich*, 20. In a 1936 commentary on the Nuremberg Laws, senior German legal bureaucrat Ernst Brandis similarly explained that the Germans were not a "unitary race, but rather one composed of various groups (Nordic, Phalian, Dinaric, Alpine, Mediterranean, East-Elbian) and mixtures between these." The "blood of all of these races and their mixtures," he explained, was that found in the German people, and therefore that "which represented German blood." Brandis, *Die Ehegesetze von 1935*, 33.

the influences of environment. Even when the tails of mice were cut off for twenty-two successive generations, the Nazi primer pointed out, their offspring were still born with them.[726] Similarly, the bloodstream of a people (race) could only be defiled by "mixing with blood" racially foreign to it.[727] In this instance, the authors could have referred to studies of blood and race—which repeatedly mentioned how blood type distributions remained consistent unless miscegenation occurred—but did not. Consequently, all German youth were instructed to recognize race predominantly in appearance, sometimes in behavior, but not in blood.

Seroanthropology was also overlooked by Nazi racial pedagogues when instructing more mature students and race theorists. Readings from the Race and Settlement Main Office of the SS (SS Rasse- und Siedlungshauptamt, or RuSHA), responsible for protecting the "racial purity of the SS," those with the "best German blood," included works by Hitler, Darré, and Günther, but nowhere was blood referenced literally as an indicator of race.[728] Nor was any such mention made in a 1941–1942 series of lectures given to the SS on race theory.[729] The SS "leadership magazine" (SS-Leithefte) referred to blood, but not in a scientific context: in 1936, the first four sections were devoted to the areas of "Blutsgedanke" (notions of blood): Peasantry, Jewry, Freemasonry, and Bolshevism.[730] The main emphasis, again, remained on physiognomy. An important element in the publication was photos and illustrations of "Aryans": blond- and blue-eyed children, and pictures of "racial adversaries" (rassischen Gegner).[731] Nonetheless, blood rhetoric remained a mainstay of SS indoctrination. According to the party line, the SS was expected not only to oversee the care and selection of "German blood" but to increase it through reproduction as well.[732] The notion of blood was also used to motivate the men in their wartime duties. Shortly after the start of World War II in September 1939, an

---

[726] Wegner, Anti-Semitism and Schooling under the Third Reich, 55.

[727] Childs, Harwood L., trans. The Nazi Primer: Official Handbook for Schooling the Hitler Youth (New York and London: Harper and Bros. Publishers, 1938), 62.

[728] NS2/152 RuSHA, April 23, 1934, and Isabel Heinemann, Rasse, Siedlung, deutsches Blut: Das Rasse- und Siedlungshauptamt der SS und die rassenpolitische Neuordnung Europas (Göttingen: Wallstein Verlag, 2003), 86–87.

[729] R/58 7353, 37. The series on "Rassenkunde" lectures was given by Clauss and Günther.

[730] Heinemann, "Rasse, Siedlung, deutsches Blut," 94.

[731] Ibid.

[732] NS 2/152, 83.

order issued from Heinrich Himmler explained that every war resulted in a loss of the "best blood." The death of the nation's best men was unfortunate but necessary; war was a "blood sacrifice."[733]

Further evidence of the irrelevance of blood type was apparent in the "race card" filled out for members of the SS, which detailed individual height, length of torso, stature, and relative length of legs, as well as the shape of the skull (very round, round, medium, long, very long) and back of the head, facial shape, nose shape (bridge, nostrils, width), cheeks, eye placement, shape of eye opening, eye folds, epicanthus (epicanthic eye fold), thickness of the lips, chin profile, hair texture, body hair, color of hair, eyes, and skin.[734] The objective was to exclude "non-European" blood, though blood typing was not part of the qualification test.[735] The men's blood would eventually be typed, recorded, and even tattooed on their arm in the event a transfusion was needed. Because of their particularly "valuable blood," members of the Waffen-SS were required to be tattooed—though this was not consistently enforced.[736]

In the interests of career advancement and siding with the view of the majority, or perhaps due to genuine disappointment and indifference, Otto Reche also abandoned seroanthropology. During the Third Reich, he continued his duties at Leipzig University's Institute for Race and Ethnology (*Institut für Rassen- und Völkerkunde*), and was further appointed as an advisor to the Race and Settlement Main Office of the SS.[737] Reche benefited from the needs of the Nazi state through his appointment as a "racial expert," whose services became all the more necessary under the Nuremberg legislation. In August 1939, due to what was described as an "overload of requests" for both genetic and racial reports (i.e., including paternity disputes and candidates for sterilization), the Anthropological Institute at the University of Munich explained that it could not for the time being accept any more.[738] The requests continued. In October

---

[733] NS 6/329, 231. From the Reichsführer-SS and Chief of the German Police in the Reich Ministry of the Interior to all SS and Police, October 28, 1939.

[734] NS2/161, 6–16.

[735] See Heinemann, *Rasse, Siedlung, deutsches Blut*, 61.

[736] NS33–167, 38–39. After the war, because of the role of the Waffen-SS in perpetrating atrocities, Allies looked for these tattoos. For this reason, many men purposely removed the tattoo, leaving an often equally suspicious scar.

[737] Schafft, *From Racism to Genocide*, 124.

[738] R 3001/20487, 449–450. To the Regional Court Munich I, August 18, 1939.

1940 the Supreme Court in Naumburg distributed a list of institutes that could assist: the Kaiser Wilhelm Institute in Berlin-Dahlem, the Institute for Genetic and Racial Research in Giessen, the University Institute for Hereditary Biology and Racial Hygiene in Frankfurt am Main, the Racial Biological Institute of the University of Königsberg, the Racial Biological Institute at Hanseatic University, and the Anthropological Institute at the University of Breslau.[739] In genetic and racial analyses, experts relied primarily on visible racial characteristics. This also seemed to apply to Reche, whose blatant disregard for the racial significance of blood was particularly conspicuous in one assignment.

In 1940, while preparing a racial analysis, Fritz Lenz wrote to the district court in Köln explaining that he could not issue an expert opinion without serological tests.[740] The Ministry of Justice became involved and referred the matter to Lenz's colleagues. Reche, Ferdinand Claussen, Eugen Fischer, Wilhelm Gieseler, and Theodor Mollison replied that they did not share Professor Lenz's views at all and that they had no difficulty in making reliable racial assessments *on the basis of photographs alone.*[741] Reche's high hopes of identifying a connection between blood and race, which had been the motivation for his establishment of the German Institute for Blood Group Research, had not been realized. Reche was now certainly aware not only that no such correlation existed, but also that seroanthropology did not impress the state or his Nazi colleagues as reliable, much less necessary. Insisting upon blood testing for racial analysis might have been considered poor judgment when the majority of racial experts favored appearance alone in determining race.

It is unlikely that Reche modified his behavior solely to align himself with the state. He had depended on physiognomy for racial classification throughout his career. Even after he was transferred to work as a racial expert in the German-occupied Eastern territories, where he could act with somewhat greater independence, Reche still did not refer to seroanthropology. After the war began, Reche became a member of the SS and was involved in "Germanizing" the East in his role as an adviser to the Race and Settlement Main

---

[739] R3001/20487, 466.
[740] Müller-Hill, *Murderous Science*, 40.
[741] Ibid.

Office of the SS.[742] Reche fit the profile of an SS physician well. On September 24, 1939, he authored a scientific expert memorandum describing methods to prevent "bastardization" and to advance ethnic cleansing in Eastern Europe.[743] He became actively involved in screening the resident population of the annexed territories in the Warthegau (occupied Poland) for "ethnic Germans," or *Volksdeutsche*, a process that was initiated in October 1939— only a month after the war began. The various categories into which individuals were placed were, like the Nuremberg Laws, based on "differences in blood." "Ethnic Germans" were entered into the "German Ethnic Register" (*Deutsche Volksliste*, or DVL), which sought to define ways of separating Germans from Poles using Nazi notions of blood. The stated purpose of the DVL was to "reclaim for Germandom" those considered to be of German blood regardless of their chosen national identity.[744] (See Figure 18.) Odilo Globocnik, a senior SS commander in occupied Poland, referred to his 1941 plan for Germanization of the General Government as "the search (*Fahndung*) for German blood."[745] According to Arthur Greiser, Gauleiter of the Warthegau, the DVL was an instrument that would maintain the memory of the pre-1939 *Volkstumskampf* (national struggle) and demonstrate that the Nazi regime was not setting out to Germanize those of "alien blood." It was the task of the racial experts to carefully scrutinize their subjects using, ironically, many of the same criteria that had been referred to in seroanthropological studies of native Germans during the Weimar Republic. Thus surnames, language, schooling, and religion were commonly referred to.

Many German anthropologists and physicians were stationed in the East specifically to decide which locals had "German blood," though there is no indication that the subjects' blood was tested for this purpose. The disconnect between metaphor and clinical reality, already apparent in the Nuremberg Laws, was now applied to the Nazis' wartime objectives. In fact, as was the case in Germany proper (the *Altreich*), "racial experts" in the East relied strongly on appearance when assessing race. Himmler repeatedly reminded researchers to take photographs of the individuals in

---

[742] Stocking, ed., *Bones, Bodies, Behavior*, 160.
[743] Michael Berenbaum and Abraham J. Peck, *The Holocaust and History: The Known, the Unknown, the Disputed, and the Reexamined* (Bloomington: Indiana University Press, 2002), 121.
[744] Harvey, *Women and the Nazi East*, 84.
[745] Heinemann, *Rasse, Siedlung, deutsches Blut*, 381.

question, to prevent the inclusion of any *"falschen Blutes"* (literally "wrong" or "non-German" blood).[746] Physical examinations were promoted, despite the size of the task at hand—the Germans planned to "resettle" (*umsiedeln*) up to thirty-one million "undesirables" from Eastern Europe to far regions of the Soviet Union.[747] After the war, Polish physician Josef Rembacz recalled in his own examination that German examiners had recorded his "height, weight, body type, skull shape, facial width, size and placement of eyes, eye and hair color, as well as hair texture"—importantly, all visible traits.[748] Reche similarly explained that it was necessary to "train the eye" in determining race, and printed high-quality illustrations and photographs to instruct others in doing so.[749] He more or less used the standard model popularized by Hans Günther and claimed that the "long-skulled" were indigenous to Europe and the "round-skulled" to Asia.[750]

Günther preferred the conventional genealogical and physiognomic criteria when determining race, though he also relied upon metaphors of blood.[751] His work *Adel und Rasse* (Nobility and Race) argued that, throughout the centuries, all the noble clans and lineages of every nation had been defined by their "Nordic blood."[752] Proof of this nobility was even apparent in "Aryan blue blood," which "only made sense in the context of the pale white skin characteristic of the Nordic race, through which the blue blood becomes visible."[753] At the same time, however, Günther denied that blood science could be used to identify racial type. Instead, he relied overwhelmingly on appearance in his racial analyses and maintained that all conceptions of human beauty, beginning with the ancient Greeks, were unconscious representations of the physical features that, from time immemorial, had been manifest in, and hence associated with, persons of noble character.[754] This was consistent with his earlier work in race theory, includ-

---

[746] Ibid., 270.

[747] Ibid., 161.

[748] Ibid., 285.

[749] Hutton, *Race and the Third Reich*, 188–190.

[750] As developed in Otto Reche, *Die Rassen des deutschen Volkes* (Leipzig: F.E. Wachsmuth, 1933).

[751] Baader, "Blutgruppenforschung im Nationalsozialismus," 342.

[752] Hans F.K. Günther, *Adel und Rasse* (Munich: J.F. Lehmann, 1926), 72–74.

[753] Günther, *Rassenkunde des deutschen Volkes*, 56.

[754] Richard T. Gray, *About Face: German Physiognomic Thought from Lavater to Auschwitz* (Detroit: Wayne State University Press, 2004), 237–238.

ing his *Racial Study of the Jewish People* (*Rassenkunde des jüdischen Volkes*). Published in 1933, this work integrated over 300 illustrations of Jews from various countries.[755] The text argued that Jews possessed unique physical qualities that marked them as an inferior race; skull measurements were used to support certain assumptions about racial classification.[756]

Reche took advantage of his wartime post to seek out—yet again—a more effective racial indicator. Previously, in his career, he had researched the nasal index, then the blood types; each time he had been seeking a potential singular, objective means of determining race. Neither of these met his expectations. Then, in April 1940, Reche expressed interest to colleague Michael Hesch in a "previously unexplored" potential racial trait—Jews' ears. Reche explained that "many observations" had been made on the shape of the Jewish nose, as well as their "cushy, bulging lips," but no research had yet been conducted on their ears. With this in mind, Reche asked Hesch:

> Photograph as many Jews' ears as possible, so that the details are recognizable. I have recently received a series of photographs of Jewish faces, and it seems to me that certain characteristics of their ears occur much more frequently than they do among the European people. I would like to further examine this issue and publish a study on Jewish ears.[757]

Reche was excited about this new prospect—so much so that he asked Hesch to keep quiet as, otherwise, someone else might attempt to "steal the idea."[758] In the event that someone did inquire, Reche further instructed Hesch to claim that the photos were being taken as references for the racial experts at the Reich Office for Genealogical Research (*Reichssippenamt*).[759]

## CLINICAL SEROLOGY

Predictably, racial studies of blood were abandoned, but non-racial applications of blood science flourished in the German courts. During the Weimar Republic, courts had recognized the ability of blood types to exclude

---

[755] Wegner, *Anti-Semitism and Schooling under the Third Reich*, 14
[756] Ibid.
[757] Geisenhainer, *Rasse ist Schicksal*, 368.
[758] Ibid.
[759] Ibid.

a potential father in cases of disputed paternity, and this practice continued uninterrupted into the Third Reich. After a series of discussions regarding the nature of the blood types and the reliability of their inheritance, the Reich Health Council explained that

> the experts are of the opinion that blood group determinations are a reliable examination which, for legal purposes—for the examination of traces of blood and for the exclusion of paternity—can be used. In some cases, identification of blood type can help decide a case which would have remained unsolved through other procedures.[760]

The Reich Ministry of the Interior dismissed lingering doubts concerning the dependability of the blood types, based in supposed exceptions to the hereditary rule. These were believed to be simply the result of erroneous blood group statistics in family trees from the 1920s. By 1939 expert serologists examined 10,000–20,000 mother-child pairs a year, and these reliably confirmed the pattern of blood type inheritance. This consistency led to their continued recognition in a court of law. The facts were so well known, the ministry declared, that cases in which the mother blatantly lied about potential fathers occasionally resulted in charges of perjury.[761] One of the tasks of the Reich Genealogical Office was to issue requests for paternity assessments. Physician reports served as a basis for decisions on "Aryan" or "non-Aryan" descent of extramarital children, children of adultery, "foundlings," and so on.[762]

In April 1939 the Reich Ministry of the Interior addressed the increasing number of illegitimate children, which was a concern to Reich authorities. It was considered important for both the "reputation and well-being" of a child to legally recognize a father. The child's fate rested on financial support, and biological fathers during the Third Reich were conventionally sentenced to pay alimony for sixteen years.[763] Fear of not being able to fulfill this obligation would drive some men, it was said, to "suicide or flight to a foreign country," while other men abandoned their regular job, explaining that

---

[760] R 3001/20487, 267.

[761] E 151/54, 70.

[762] Maria Teschler-Nicola, "*Volksdeutsche* and Racial Anthropology in Interwar Vienna: The 'Marienfeld Project,'" in Turda and Weindling, eds., *Blood and Homeland*, 70.

[763] E 151/54, 377.

culosis, venereal disease, malaria, asthma, or any disorders of the heart, blood, or nervous system. In addition, all drinkers and smokers were to be declined.[781] Usually, women were not to give blood while menstruating or pregnant. Aside from the nominal requirement that the donor be "of German or Aryan descent," race did not factor into the procedure of exchanging blood. Seroanthropology was not referenced; type B was not flagged as "non-German"; and no proteins or chemical reactions were used to confirm the racial type of the donor.

Germany was progressive in its legal recognition of blood science, but similar advances were not made in the area of transfusion therapy. By not storing blood and accepting only "Aryan" donors-on-the-hoof, the Germans severely restricted their blood supply. The Allied powers, by contrast, had already established regular systems of blood donation, banking, and transfusion in the decade following World War I.[782] In spite of the millions of casualties in that war, doctors had only transfused perhaps a few hundred, but the implications of those results led to further research.[783] (See Figure 19.) By the late 1930s the American system, largely shaped by the efforts of Karl Landsteiner, was internationally recognized for its efficiency.[784] In Germany most surgical institutes kept lists of potential donors, who would then be called as needed.[785] The organizers of a new blood transfusion facility in Gmünd during the war emphasized that donors needed to be immediately available for donation both day and night by telephone. If these individuals did not have a phone, the instructions continued, "there must be one nearby made immediately available to them."[786] The Germans did experiment with preserved blood, chiefly between 1940 and 1942, but there were so many serious reactions that medical officers apparently lost interest in it.[787] Some medical officers had never seen preserved blood

---

[781] E151/54, 378. The donors were required to undergo testing every three months and usually did not give more than 500 ccm of blood each session. Afterwards, they were not to donate again for six weeks.

[782] Starr, *Blood*, 55.

[783] Ibid.

[784] D. Wiebecke et al., "Zur Geschichte der Transfusionmedizin in der ersten Hälfte des 20 Jahrhunderts: unter besonderer Berücksichtigung ihrer Entwicklung in Deutschland," *Transfusion Medicine and Hemotherapy* 31 (2004): 22.

[785] E. von Rosztóczy, "Untersuchungen über Isohämagglutination in der Umgebung von Szegedin," *Zeitschrift für Rassenphysiologie* 4 (1931): 145.

[786] E 151/54/ 378.

[787] Douglas B. Kendrick, *Blood Program in World War II* (Washington, DC: U.S. Government Printing Office, 1964), 22.

used in the field without "deleterious chills."[788] Plasma and serum were seldom used, although officers who used captured U.S. stocks of plasma were enthusiastic about it.[789]

No system of storing blood comparable to that in the United States, Russia, or Britain was established in Germany. In fact, German transfusion therapy lagged well behind these nations since 1933, which led to the German military being woefully under-supplied with blood during the war. Instead of bottled blood, selected soldiers were often appointed to travel to the front with sanitary units to give blood.[790] The Germans even entertained the idea of using cadaver blood for transfusions, as one physician optimistically reported that blood could be drawn from a corpse for "up to eight hours after death."[791] Though the Allies likewise considered using cadaver (and placental) blood, the advancements they made in storing blood led them to disregard such possibilities. Even late in the war, the Germans were still having problems with the most basic task of typing blood. The frustration of authorities is clear in a letter issued by the Ministry of the Interior to various officials in April 1944:

> Incorrect blood type determinations are still occurring. In fact, their incidence has increased as of late. This has made it necessary, in spite of the war, to detain the authorized serologists for advanced instructional courses on performing blood typing. All authorized serologists are to take part in these courses within the next year. Despite the priorities of the war, blood typing experts are expected to devote their time to perfecting their technique, as it is obviously flawed.[792]

One of the assigned lectures in the program was on the "sources of error when typing blood." Inaccurate results could have significant legal ramifications and, more importantly, life-threatening consequences for the recipients of incompatible transfusions. Overall, German blood science was inferior to that of the Allied forces. Allied medics who came upon German field hospitals commonly found wounded soldiers bled white.[793] So lacking was

---

[788] Ibid.

[789] Ibid.

[790] Starr, *Blood*, 115.

[791] A. Mahlo, "Über Bluttransfusion," *Deutsche Medizinische Wochenschrift*, no. 49 (December 8, 1939): 1762.

[792] E 151/54, 377.

[793] Starr, *Blood*, 120.

German transfusion technology that American doctors who came upon enemy wounded often felt compelled to administer plasma.[794] The British and Americans had learned to give large quantities of blood through slow-drip transfusions, while the Germans seemed completely ignorant of this technique and gave tiny amounts of blood, fearful of overtaxing the heart.[795] This, combined with the exclusion of non-Aryan blood, caused Paul Schultze, a German field surgeon captured by the Russians in 1942, to complain about the Party's "senseless race theories" and requirement for "pure blood," which made it impossible to collect an adequate supply.[796] In early 1942 General Dr. Wolff, chief of the German department of hygiene and sanitation, emphatically declared that the medical preparations for war on the Eastern Front were totally insufficient. For this failure he blamed Dr. Leonardo Conti, Reich health leader and head of the Department of Health in the Ministry of the Interior. During the courses on medical preparedness, Conti focused more on "ideological teachings" than on practical instruction.[797] Wolff asked the proper authorities for two changes in order to improve the courses of war medicine, namely 1) discontinuance of the ideological courses of the Nazi Party with the military and racial views of the party and 2) immediate rehabilitation of Jewish physicians for service. However, Conti obtained support from the authorities for the refusal of Wolff's first request. The second request was only partially accepted: Jewish physicians were rehabilitated for service as auxiliary physicians in civil hospitals but not in the army.[798]

The German medical establishment, as well as the general public, were relatively unfamiliar with blood transfusion therapy. Due to this inexperience, the German people were often reluctant to donate blood. The Ministry of the Interior estimated that one donor was needed for every 1,000 citizens, which meant that approximately 60,000 donors were required in Germany.[799] A 1940 report to Württemberg authorities noted that only three or four donations had been collected the previous year in the

---

794 Ibid., 115.
795 Ibid.
796 Ibid.
797 "Failure of German Army Medical Service," *Journal of the American Medical Association* 119, no. 8 (May 1942): 665.
798 Ibid.
799 "Richtlinien für die Einrichtung des Blutspenderwesens im Deutschen Reich," *Ministerial-Blatt des Reichs- und Preussischen Ministeriums des Innern* 5, no. 11 (1940): 449.

regional hospital.[800] Despite extensive advertising in newspapers and factories, the health office in Backnang complained that it had only thirteen available donors.[801] In 1940 the general hospital in Ebingen reported that it could not support a blood bank due to a shortage of donors.[802] In 1940 *Neues Volk* reported that "countless men" had already been rescued through blood transfusions, but the publication added that the number of donors was still not enough and pleaded with the populace to donate blood regularly and serve as givers of life to the *Volksgemeinschaft*.[803] The article reassured that donating was not dangerous for the donor; giving and receiving blood was conducted "quickly and painlessly."[804] No such pleading was necessary in the United States, Great Britain, or the USSR, where people rushed to donate in large numbers. For a period in 1942, the American Red Cross actually exceeded the military's blood quota.[805] The following year, the quota rose to more than a million pints, but this too was met.[806] In Britain during the war, thirteen million units were collected over five years from a population of forty million people—approximately 30 percent higher than the modern donation rate (5 percent of the population per year).[807] In Moscow, more than 2,000 citizens lined up each day to give blood.[808] Blood drawn was used not only as whole blood but also for the production of albumin and freeze-dried plasma.

Having killed or expelled their most able physicians, and rejecting foreign medical advances, the Germans relied on medical technologies at least a generation out of date.[809] After 1933 the majority of Jewish doctors in Germany and German-occupied territories were forced out of their profession. Many of them perished in the Holocaust. Most, however—an estimated 9,000 to 10,000 German-speaking doctors—succeeded in emigrating, the majority of them to the United States.[810] Under the Nuremberg Laws,

---

[800]  E 151/54, 378.

[801]  Ibid.

[802]  Ibid.

[803]  *Neues Volk* (1940): caption 26.

[804]  Ibid., caption 27.

[805]  Starr, *Blood*, 110.

[806]  Ibid.

[807]  J.R. Hess and M.J.G. Thomas, "Blood Use in War and Disaster: Lessons from the Last Century," *Transfusion* 43 (2003): 1630.

[808]  Starr, *Blood*, 117.

[809]  Ibid., 115.

[810]  Natalia Berger, ed. *Jews and Medicine: Religion, Culture, Science* (Philadelphia and Jerusalem: The Jewish Publication Society, 1995), 158.

# CONCLUSION

While scientific studies of blood and race were not utilized by the Nazis in enforcing the identification and separation of different racial types, blood rhetoric remained politically useful. German race ideologues were able to exploit the flexible notion of blood defilement, which could function in either a clinical or sexual context. As advancements were made in blood science, these were incorporated into racial propaganda because science was felt to project an "objective and value free" image.[824] The competing political ideologies of the interwar period were evident in studies of blood, which were related to widely disparate areas of concern. During the Weimar Republic, a paradigm was established whereby people were divided into those possessing "value" and those lacking it.[825] Through analyses of groups "lacking in value," such as criminals, asylum inmates, and diseased persons, researchers attempted to apply blood science to the "falling birthrate, corruption, and various forms of degeneration" of the time.[826] Infectious disease and mental instability were traits perceived as disadvantageous to the *Volkskörper*'s well-being. Researchers hoped that blood science might assist in distinguishing and protecting the purity of German blood. In spite of the intriguing patterns occasionally observed, comparisons between blood type and other characteristics lacked precision. For purposes of racial classification, seroanthropology appeared to have the same disadvantage as physical anthropology in that no one feature could be used to determine

---

[824] Efron, *Defenders of the Race*, 2.
[825] Peukert, "The Genesis of the 'Final Solution,'" in Thomas Childers and Jane Caplan, *Reevaluating the Third Reich* (Teaneck, NJ: Holmes and Meier, 1993), 277.
[826] Proctor, *Racial Hygiene*, 305.

race. Examiners who were convinced that there were different racial types, and that each had its place in a racial hierarchy, made a much more concerted effort than their unbiased colleagues to determine the racial significance of blood. Even with its best efforts, however, the German Institute for Blood Group Research failed to achieve its main objective of proving a relationship between blood and race. Statistics that initially held promise were gradually chipped away at; repeated research confirmed that there was no link between blood type and race, physiognomy, mental disorders, or illness. The fledgling science of seroanthropology was doomed by the fact that there was no association between blood and race. Any connections that had been made crumbled under close medical scrutiny. Seroanthropologists became increasingly pessimistic as inconsistencies and contradictions grew ever more conspicuous.

For *völkisch* scientists, arguably the most committed of the lot, their patience wore thinner as not only unpredictable, but *unwanted* results turned up. There were statistics to suggest that extensive miscegenation had occurred between "Aryan and non-Aryan types," resulting in *Mischlinge* with "mongrel blood." This especially seemed to be the case in eastern Germany, where levels of "Western" type A blood were generally much lower. Disparities in distributions of blood type within Germany contradicted National Socialist claims of "pure blooded" Germans. The medical evidence, the typing of the blood of thousands of Germans, had literally proved this wrong. What is interesting is that seroanthropology was much more likely to be rejected by race scientists during the Third Reich—even though other methods of racial classification were similarly problematic. Both Nazi "racial experts" and bureaucrats tended to prefer using traditional physiognomic racial indicators, even as they occasionally criticized this approach. Anthropomorphic characteristics, particularly the shape of the skull, had been the prevailing means of racial differentiation long before the Nazis. In the late nineteenth century, German anthropologists arranged for the collection of murdered natives' skulls from German Southwest Africa for racial research. Later, Weimar scientists hosted the "Best Nordic Head" contest. Both serve as good examples of a preference that spanned many decades and was carried into Nazism. Immediately after *Kristallnacht*, university and museum anthropologists rushed to a stadium where Jews had been rounded up and interned, to make plaster-cast masks of hundreds of prisoners.[827] Haste was necessary, for these prisoners were being deported to Buchenwald, and many would not return.[828] National Socialists were merely being consistent in what was by then a very familiar emphasis on appearance. Mere observation of the different "shapes and colors" of the races was certainly easier to present to the German public—and, therefore, much more effective in race propaganda—than the frequencies of blood type distribution and their accompanying biochemical indices.

Those who had been schooled in the racially charged intellectual milieu of early-twentieth-century Germany, such as Otto Reche and Ludwik Hirszfeld, were accustomed to racial differentiation through physiognomy. However, they were also conscious of its drawbacks. The unreliable nature of visible traits led both Hirszfeld, and eventually Reche, to pursue more dependable methods of identifying race. Eventually, Hirszfeld did recognize that his work had introduced a "discourse of segregation."[829] Because it assigned Europeans the highest biochemical index, and Asians and Africans the lowest, Hirszfeld's theory of serological racial types preserved a racial hierarchy, whether intentional or not.[830] One author claims that the biochemical index, with its scale putting Northern and Western Europeans at the top, served "the same pseudoscientific purposes that the cephalic index had for Aryan racists."[831] It had not, however, been Hirszfeld's intent to reinforce the idea of "superior" and "inferior" races. His contribution was shaped by the race-conscious environment of the German medical establishment in which he was schooled and practiced. In his work on Jewish race scientists in fin-de-siècle Germany, John Efron claims that Jews took part in racial discourse mainly to respond to the scientific, and in truth often racist, discourse on human variation. A crucial feature of these Jewish scientists was their complete rejection of racial science for chauvinistic purposes—in spite of their use of the contemporary language, and methodology, of race science.[832] This also applied to Ludwik Hirszfeld who, like the vast majority of educated whites in Europe and North America of his day, accepted the idea that races existed and that they fundamentally differed

---

[827] Schafft, *From Racism to Genocide*, 101.

[828] Ibid.

[829] Dickinson, "Biopolitics, Fascism, Democracy: Some Reflections on Our Discourse About 'Modernity,'" 22.

[830] Schneider, *Quality and Quantity*, 221.

[831] Ibid., 228.

[832] Efron, *Defenders of the Race*, 9.

from one another both mentally and physically.[833] It was simply under-stood that all peoples could be grouped taxonomically. Historian Sheila Faith Weiss has pointed out that, from today's vantage point, all German eugenicists would be considered racist, and this type of racism was shared by most eugenicists everywhere.[834] The study of racial anthropology was not considered "aberrant or 'pseudoscientific' by the international scien-tific community," as Robert Proctor notes.[835] As "good" acculturated Euro-peans, Jewish physicians could not help but believe in the concept of race.[836]

This explains why Jewish physicians, or those of Jewish descent, sought to contribute to the intellectual study of races and assumed the language common to race scientists, regardless of their political affiliation. Doing so did not make them sympathetic to Nordicism, much less place them in the same camp as *völkisch* race scientists. Because the articles were well-researched and applicable to scientific studies of race, even the far-right German publication *Politisch-Anthropologische Revue* reported periodically (and without criticism) news from Jewish publications such as *Ben Israel* and the *Jüdisches Volksblatt*.[837] For the same reason, the editors of the peri-odical of the German Institute for Blood Group Research accepted articles by Jewish authors and invited others of Jewish heritage to join the institute.

In 1938 Hirszfeld had specifically denounced political appropriations of medicine that were to reach unimaginable heights during the war. As a converted Jew, Hirszfeld was affected by the German invasion of Poland. The fact that Hirszfeld was a practicing Roman Catholic was irrelevant to the Nazis, who categorized him as (racially) Jewish. In February 1941, Hirszfeld and his wife and daughter, along with hundreds of thousands of Jews from Warsaw and other towns in Poland, were herded into the Warsaw Ghetto.[838] When deportations began in earnest and the ghetto hospital was liquidated, Hirszfeld managed to escape to the "Aryan" side of Warsaw with his family.[839] They eventually found refuge in the countryside and were

---

[833] Ibid., 176.

[834] Weiss, "The Race Hygiene Movement in Germany," 194.

[835] Robert Proctor, "From Anthropology to *Rassenkunde* in the German Anthropological Tradition," in Stocking, ed., *Bones, Bodies, and Behavior*, 140.

[836] Efron, *Defenders of the Race*, 176.

[837] Proctor, *Racial Hygiene*, 142.

[838] Heynick, *Jews and Medicine*, 443.

[839] Ibid.

hiding in the woods when the Russian army came through. Hirszfeld and his wife Hanna survived, but their only child, Maria, died of pneumonia. As the distraught Hirszfeld remarked, at least she had been buried in her own grave, even if under a false name.[840] After the war, Hirszfeld resumed his work in Warsaw, where he passed away in 1954.

In stark contrast to Hirszfeld's unbiased research, Otto Reche's involvement in seroanthropology was motivated by a political mix of Aryan supremacy, pan-Germanism, and anti-Semitism. Reche's foray into seroanthropology was the most significant divergence in his career from physical anthropology, though his attitude towards blood science gradually shifted from intermittent doubt to complete rejection. Eventually, even Reche admitted that having type B blood was in no way an "indicator of Jewishness."[841] National Socialists were also aware of this, as were the leading race theorists of the day. At the same time, the Nazis used "blood" as a marker of racial difference—a metaphor previously limited to propaganda. Objectively articulating this notion, and protecting "German blood" from defilement in the sense demanded by the Nuremberg Laws, often resulted in "absurdity piled on top of absurdity."[842] Blood "defilement," or "pollution," was believed possible from literal exchanges of blood or sexual contact. In prohibiting blood transfusions between Germans and Jews, the Nazis were applying the ancient concept of "vitalism," the belief that blood somehow carried the essence of the creatures (or persons) in which it flowed.[843] Medical experiments and therapies over previous centuries, in which the blood of gladiators was consumed for strength, or the blood of a lamb was given to calm a hot-tempered individual, indicate this belief that blood could also transmit character traits.[844] The same idea of vitalism prompted one nineteenth-century physician to voice concerns that a transfusion of ox blood would cause a human recipient to grow horns, just as it did the Nazis to fear that Jewishness was somehow "contagious" through blood. *Völkisch* theorists were able to shape their propaganda around the notion that blood was

---

[840] Ibid., 444.

[841] Karl Saller, *Die Rassenlehre des Nationalsozialismus in Wissenschaft und Propaganda* (Darmstadt: Progress Verlag, 1961), 101.

[842] Schleunes, ed., *Legislating the Holocaust*, 19.

[843] Starr, *Blood*, 5.

[844] Roy Porter, *The Greatest Benefit to Mankind: A Medical History of Humanity* (New York: W.W. Norton and Company, 1997), 39.

more than just a bodily fluid but instead represented the soul. Both concepts were apparent in the Serelman case, in which a Jewish physician gave his blood to save the life of an SA man. The donor's blood was Jewish, but also "masculine" because of his service in World War I, and therefore not considered a threat. This justification demonstrates well the flexibility of the Nazi concept of race, defined by Hans Günther as "a group of individuals differentiated from all others by a unique combination of bodily and spiritual characteristics."[845] This allowed the Nazis to incorporate a convenient mix of physiognomic and psychological traits when determining race. Appearance may have been the preferred means of differentiation, but it was not definitive, as demonstrated in the anti-Semitic novel *The Operated Jew*, in which the subject looked German but lacked a "German soul"—an integral part of belonging to the German race. Nazi "racial indicators" varied precisely because no one characteristic was unique to any one race.

While the Third Reich has often been criticized for suppression of medical intellect and progress, certain sciences actually flourished—namely "psychology, anthropology, human genetics, and various forms of racial science and racial hygiene."[846] Racial studies of blood did not prosper, however, and seroanthropology did not play an important role in the modern movement in science to discover a solution to Germany's "racial problems" by "naming, defining, measuring, quantifying, and investigating."[847] In addition, Nazi propagandistic use of blood rhetoric had very little to do with the scientific study of blood. The course of seroanthropology, and its ultimate neglect during the Third Reich, cast serious doubt on the current "biopolitical" account of modern German history, which describes the Nazi racial state as a phenomenon with identifiable and direct historical roots in the sciences of pre-Nazi Germany.[848] From one perspective, it is easy to sense impending disaster in pre-1933 German medicine, especially in the areas of eugenics and anthropology.[849] There is no comparable line of continuity between Weimar studies of blood and race and the atrocities later perpetrated by the Nazis. Seroanthropology effectively contradicts

---

[845] As quoted from Günther's *Rassenkunde*, 14.

[846] Proctor, *Racial Hygiene*, 5.

[847] Dickinson, "Biopolitics, Fascism, Democracy: Some Reflections on Our Discourse About 'Modernity,'" 2.

[848] Ibid., 5.

[849] Ibid.

the notion that interwar biopolitics was saddled with such negative, even lethal, potential. Because they were not disastrous, studies of blood and race represent a fundamental divergence from this narrative. They met with increasing disregard during the Weimar Republic and were found to be inadequate for the Nazis' purposes of individual racial classification. When considered in the vast expanse of race science, seroanthropology had relatively few enthusiasts; it was barely acknowledged, much less relied upon, by the leading race theorists of Weimar and Nazi Germany. Particularly telling is the fact that the Kaiser Wilhelm Institute in Berlin, considered the leading institute of racial anthropology during the Third Reich, issued no publications in which ethnic distribution according to blood type was the principal subject.[850]

After the war, medical organizations associated with the Nazi regime were either disbanded or outlawed.[851] No such measures were necessary with seroanthropology, however, as it faded into obscurity well before the war's end. The German Institute for Blood Group Research closed in 1944, and its periodical, the *Zeitschrift für Rassenphysiologie*, stopped publication in 1943. In its final years, issues of the journal were substantially slimmer than they had been at the institute's height in the 1920s.

Similarly, the leading *völkisch* German blood scientists faded into insignificance after the war. There was no need to "make an example of" seroanthropologists as there had been with other physicians, because their work was not believed to have been involved in constructing the murderous policies of the Third Reich. On April 16, 1945, Otto Reche was arrested by American forces and relocated to Hamburg for denazification, but was released after only sixteen months of being detained.[852] He continued his work in anthropology, genetics, and paternity testing until his death in

---

[850] Benoit Massin, "Rasse und Vererbung als Beruf: Die Hauptforschungschrichtungen am Kaiser-Wilhelm-Institute für Antropologie, menschliche Erblehre und Eugenik im Nationalsozialismus," in Hans-Walter Schmuhl, ed., *Rassenforschung an Kaiser-Wilhelm-Instituten vor und nach 1933* (Göttingen: Wallstein Verlag, 2003), 212–213. In 1933–1934, Spanish guest scholar Jimena Fernández de la Vega was occupied with measuring the size of the blood corpuscles of twin subjects, but no publication emerged. Serological research was also conducted by Engelhard Bühler. His work on the heritability of antibodies in the blood was presented at the Convention of the German Society for Genetics in Jena in 1935. Eugen Fischer did apply to the Reich Research Council for funds to continue Bühler's research. Fischer gave two reasons for his interest—the possible relation of the clumping characteristics to race, and their potential to improve immunizations against infectious disease. See Schmuhl, *The Kaiser Wilhelm Institute*, 175–176.

[851] Proctor, *Racial Hygiene*, 299.

[852] Geisenhainer, *Rasse ist Schicksal*, 487.

March 1966. Paul Steffan, Reche's colleague and co-founder of the institute, served as an expert serologist for the courts after the war. He was also appointed director of the Office for Forensic Blood Type Research with the regional court in Berlin, as well as the laboratory for serological and blood type research in Nikolassee.[853] Unfortunately, the Allies were too lenient in their postwar prosecution of Nazi physicians. Sources indicated that von Verschuer continued his research of "specific proteins" in the blood until the last phases of the war. With the assistance of his colleague Josef Mengele at Auschwitz, he coordinated the shipment of both blood and tissue samples from camp inmates. In October 1944, with the Russians drawing closer to Auschwitz, Mengele still wrote a follow-up report to von Verschuer in Berlin: "Further research is being carried out with Hillmann" (a colleague from the Kaiser Wilhelm Institute of Biochemistry).[854] Over the course of their project, Mengele traveled to Berlin several times to report to von Verschuer on the progress of their work. Even as late as January 6, 1945, von Verschuer wrote to a colleague, "you will be interested to know that my research on specific proteins has finally reached a decisive stage now that we have overcome some considerable methodological difficulties."[855] A few days later Mengele left Auschwitz, and the circumstances of war prevented von Verschuer from finishing his work, though he likely had time to destroy what was apparently the most damning evidence of it.[856] After the war, von Verschuer was judged a collaborator (*Mitläufer*) and hence absolved from any major responsibility in the events of the Nazi period; he was fined 600 marks and released from custody.[857] Von Verschuer was exonerated in 1949 and in the early 1950s was both correspondent for the American Society of Human Genetics and a member of the Royal Medical Society in London.[858]

---

[853] Ibid.

[854] Hillmann was working in Professor Butenandt's institute in Dahlem on a scholarship from the Reich Research Council, although Professor Butenandt himself had just moved to Tübingen. See Müller-Hill, *Murderous Science*, 78.

[855] Ibid.

[856] Dr. Mengele's letters and reports were probably destroyed by von Verschuer. See ibid., 79.

[857] Ibid.

[858] Anne Cottebrune, "The Deutsche Forschungsgemeinschaft (German Research Fund) and the 'Backwardness' of German Human Genetics after World War II," in Wolfgang Uwe Eckart, ed., *Man, Medicine, and the State: The Human Body as an Object of Government-sponsored Medical Research in the 20th Century* (Stuttgart: Franz Steiner Verlag, 2006), 99.

While no Germans were prosecuted and hanged after the war for clinical studies of blood and race, one prominent Nazi was executed for the promulgation of virulently anti-Semitic propaganda that made frequent use of blood rhetoric. Julius Streicher, founder of the Nazi publication *Der Stürmer*, was found guilty by the Nuremberg tribunal of crimes against humanity and hanged on October 16, 1946. However, unlike the other defendants, Streicher had never been personally involved in carrying out Nazi policies. It became clear that he was notorious for his rhetoric rather than for any actions taken. Nonetheless, he was assigned a seat on "murderer's row" in the dock at the International War Crimes Tribunal. The tribunal referred to his writing and publications as "propaganda of death."[859] Streicher was most renowned for his work as editor-in-chief of *Der Stürmer* from 1923 to 1945, in which "blood" was a recurring theme. Ridiculous claims were made concerning differences in blood, with references to Jewish blood being "animal" in origin, or toxic if transfused into an "Aryan." Lurid descriptions of "blood defilement" and photographs of "blood defilers" were featured throughout. Streicher's use of blood rhetoric extended to his political causes as well; he proposed classifying as Jews individuals with "one drop" of Jewish blood instead of the larger proportion allowed by the Nuremberg Laws.[860] Such rhetoric was considered lethal by the Nuremberg judges, and hate speech—not actions—was prosecuted as a war crime. Streicher was hanged even though he had held no official government or party position and *Der Stürmer* was not an official party or state organ.[861]

As indicated by the volume of research during the interwar period on blood type and race, however, Germany was unique in its pursuit of seroanthropology, though fervent interest was also expressed in Central and Eastern Europe. Besides Germany, Russia was the only nation to establish an institute dedicated in large part to determining the link between blood and race; the Ukrainian Standing Commission on Blood Group Research was established in 1927, only a year after its German counterpart. The amount of work done by Soviet and German researchers was so large and concen-

---

[859] Eric Zilmer, *The Quest for the Nazi Personality: A Psychological Investigation of Nazi War Criminals* (London: Routledge, 1995), 152.

[860] Schleunes, ed., *Legislating the Holocaust*, 153.

[861] Allan Thompson and Kofi Annan, *The Media and the Rwanda Genocide* (London: Pluto Press, 2007), 337.

trated that it distorts the overall pattern of seroanthropological research over time. Without them, the peak in the late 1920s is much lower (about half the articles in 1927 and 1928), and the decline in the 1930s is much less pronounced.[862]

German and Eastern European blood scientists regularly corresponded with one another.[863] A similar interest in blood and race never developed among Western researchers, who devoted relatively little attention to seroanthropology. In Great Britain, a total of only six articles of original research were published on blood type distribution by 1932.[864] In the United States, at least initially, many authors seemed interested only in the technicalities of blood typing and gave no references or any description of the people tested. An article by Moffitt et al. reported the results of blood tests on soldiers at Fort McPherson, Georgia. The authors sought to obtain an average representation in the population tested. "The sera we used," they boasted, "were from adults of all ages, from all classes of society, white and Negro races." Even when Culpepper and Abelson of Detroit's Parke Davis Laboratory published an article in 1921, there was no mention of Moffitt et al., let alone the Hirszfelds' work of 1919. Culpepper and Abelson simply described the distribution of the ABO types for the 5,000 subjects examined.[865] A French article of primary research was not presented until 1925, many years after the Hirszfeld study.[866] After only a brief period in the early 1920s, German quickly superseded English as the most common language of publications on blood type distribution.[867]

The source of this disparity in interests between East and West lies in the social and national changes brought about by World War I. The effects of these changes were most pronounced among the defeated powers, whose intellectual and political elite often looked to eugenic and racial principles as "a source of hope in their disillusioning environment."[868] Like its better-known cousin racial anthropology, seroanthropology was another

---

[862] Schneider, "Initial Discovery," 288.

[863] See Marius Turda, "Entangled Traditions of Race: Physical Anthropology in Hungary and Romania, 1900–1940," Focaal 58, 3 (2010): 32–46.

[864] Schneider, "Initial Discovery," 300.

[865] Ibid., 284.

[866] Schneider, Quality and Quantity, 223.

[867] Schneider, "Initial Discovery," 301.

[868] Marius Turda, "The First Debates on Eugenics in Hungary, 1910–1918," in Turda and Weindling, eds., Blood and Homeland, 204.

type of science applied to these principles. Despite a comparative indifference concerning studies of race and blood, the United States had enacted a series of laws which, like the Nuremberg Laws, also discriminated on the basis of "blood."

Virginia's Racial Integrity Act, introduced in 1924, defined race by the so-called "one-drop rule," which ruled that anyone with so much as "one drop of colored or Negro blood" would be classified as belonging to those groups.

> It shall hereafter be unlawful for any white person in this State to marry any save a white person, or a person with no other admixture of blood than white and American Indian. For the purpose of this act, the term "white person" shall apply only to the person who has no trace whatsoever of any blood other than Caucasian; but persons who have one-sixteenth or less of the blood of the American Indian and have no other non-Caucasic [sic] blood shall be deemed to be white persons.[869]

The most virulent Nazis admired and referred to this legislation, arguing that the Nuremberg Laws should have "gone further," insisting likewise that "one drop" of Jewish blood was sufficient to make any person Jewish.[870] Like their more lenient German counterparts, the Racial Integrity Act and other American laws policed miscegenation through the use of blood metaphor. In Georgia a "white" person was categorized as one who had "no ascertainable trace of Negro, African, West Indian, Asiatic Indian, Mongolian, Japanese, or Chinese blood in their veins." And Tennessee referred to "Negroes, mulattoes, or persons of mixed blood descended from a Negro to the third generation inclusive."[871] Such legislation led to a flurry of miscegenation court cases in the United States in the 1920s, but actual studies of blood and race remained exceedingly rare in spite of seroanthropology's relative popularity at this point in Europe.

Like German race theorists and ideologues, some Americans also expressed concern about the "defilement" of blood through both sexual

---

[869] Peggy Pascoe, "Miscegenation Law, Court Cases, and Ideologies of 'Race' in Twentieth-Century America," *Journal of American History* 83, no. 1 (1996): 44–69, 59.

[870] Schleunes, ed., *Legislating the Holocaust*, 2.

[871] Peggy Pascoe, *What Comes Naturally: Miscegenation Law and the Making of Race in America* (Oxford: Oxford University Press, 2009), 144 and 347.

and clinical means, even though U. S. physicians readily acknowledged that there was no difference between the blood of blacks and whites. In a 1943 article entitled "What We Do Not Know about Race," Doctor William M. Krogman explained that:

> all blood groups and their genes are found in Whites, Yellows and Blacks, though in varying percentage combinations. It is possible that these combinations may have some value in racial distinction, just as does skin color, etc., but as far as transfusability is concerned (allowing for blood groups) all human blood is alike.[872]

Charles R. Drew, an African-American physician who made critical contributions to collecting and storing the blood supply for World War II, stated to reporters, "I will not give you an opinion [on the matter of segregating blood]. I will give you the scientific facts in the matter. The blood of individual human beings may differ by blood groupings, but there is absolutely no scientific basis to indicate any difference according to race."[873]

Such reassurances from the medical community, including even public statements from scientists at the Red Cross, were common. Nonetheless, mainly because of racial segregation (including, notably, in the armed forces), the American Red Cross kept its stores of blood separate according to race. Health centers as renowned as the Johns Hopkins Hospital in Baltimore segregated blood, while others did not accept "Negro blood" at all.[874] As the war progressed, blood from black donors was accepted but was labeled and processed separately. This way, the Red Cross explained, "those receiving transfusions may be given plasma from the blood of their own race."[875] As was the case in Nazi Germany, the prohibition of interracial blood exchange prompted questions. One individual asked the editor of the *Journal of the American Medical Association* whether there was any

> definite basis in serology, pathology or hematology for the aversion to the use of [black] blood? Was knowledge sufficiently complete with reference to all

---

[872] William M. Krogman, "What We Do Not Know about Race," *Scientific Monthly* 57, no. 2 (1943): 101.

[873] Spencie Love, *One Blood: The Death and Resurrection of Charles R. Drew* (Chapel Hill: University of North Carolina Press, 1997), 321.

[874] Starr, *Blood*, 98.

[875] Ibid., 108.

the blood elements for one to be able to say without contradiction that the use of Negro blood for blood banks was 100 percent safe?[876]

The editor explained that various chemical and serological investigations into the matter, including Manoiloff's "serochemical reaction," had yielded no evidence that the blood of one race could be distinguished from that of another. The reaction had not been used to try to distinguish "Negroes from Caucasians," but such attempts had been largely addressed (and abandoned) by immunologists. Other efforts to differentiate between "white" and "black blood" had similarly failed, leading the editor to conclude that there was no "factual basis for the discrimination against the use of Negro blood or plasma for injection into white people." The aversion was, he speculated, perhaps the result of "ancient folklore" that associated one's personality with his or her blood. Interestingly, the question about using blood from black donors was posed by a physician. Clearly, misconceptions about the blood of different races lingered even within the medical establishment. In 1940 French physician Arnault Tzanck expressed his "amazement" to have injected blood from a white individual into a black "without incident."[877]

In large part because Americans were fighting a racist enemy, while simultaneously keeping a segregated blood supply, the policies of the Red Cross provoked controversy and protest from U.S. general public and medical professionals. The *New York Times* editorialized: "We cannot explain the prejudice that the Red Cross is keeping alive [and assume it] is a survival of the superstition and mysticism associated with blood... Sometimes we wonder whether this is really an age of science."[878] Public demonstrations protesting the policy were common.[879] Still, the Red Cross found it impossible to overcome the assumption that "most men" of the white race would object to receiving the "blood of Negroes."[880] Indeed, there had been reports to suggest that this was the case. Not surprisingly, accounts of German soldiers refusing blood on a racial basis were more common. Two

---

[876] "Use of Negro Blood for Blood Banks," *Journal of the American Medical Association* 119, no. 3 (May 16, 1942).

[877] Jean-Francois Picard and William H. Schneider, "L'histoire de la transfusion sanguine dans sa relation a la récherché médicale," *Vingtième Siècle* 49 (1996): 4.

[878] Starr, *Blood,* 108.

[879] Love, *One Blood,* 321.

[880] Ibid., 212.

French blood scientists stationed in North Africa during the war recounted how, in Tunis, injured Germans refused to be given blood from Italian prisoners. Unbeknownst to the Germans, however, the French physicians used bottled blood from North African Jewish and Arab donors.[881] At the same time, "white" blood was very much in demand among not only Germans, but also white American troops in North Africa, as the U.S. military refused to mix the blood of white and black donors.[882] In fact, in 1941, the U.S. Armed Forces announced that black donors would be excluded from the American Red Cross blood banks then being set up across the country.[883] Later on in the war, after many individuals and civil rights group had protested the policy of exclusion, the Red Cross adopted a new policy of accepting black donors but segregating black and white blood.[884] This was followed as official nationwide policy until 1950.[885] The catastrophe of the war and the Third Reich left German medicine very sensitive to particular topics, particularly those areas of racial anthropology and eugenics that were manipulated to serve Nazi interests.[886] This included racial studies of blood, even though seroanthropology had never been a main priority of German race theorists. In postwar Germany political circumstances dictated "practical" blood science, and the focus shifted entirely to transfusion therapy and legal applications. Correspondence from the Berlin Blood Donor Service in 1951 emphasized the importance of blood transfusions:

> For many sick or wounded, a blood transfusion is one of the most effective therapies in the hand of the physician. Its timely use commonly decides life or death for a patient. Furthermore, in the last decade, blood transfusions have experienced a considerable increase from the efforts of medical progress and practice. In the Berlin Hospital alone, 10,000–12,000 blood transfusions are annually required. This volume cannot be gathered through the previous small circle of paid blood donors. For this reason, the Berlin Blood Donor Service emphasizes the urgent need of blood- and plasma products.[887]

---

[881] Picard and Schneider, "L'histoire de la transfusion sanguine dans sa relation a la récherché médicale," 10.
[882] Ibid.
[883] Love, One Blood, 49
[884] Ibid.
[885] Ibid.
[886] Proctor, Racial Hygiene, 303.
[887] R 86/4214.

There was no comparable stigma against studying race and blood elsewhere, however, and seroanthropological research in other nations continued. In the first volume of his pivotal series on the history of anti-Semitism, originally published in 1955, author Leon Poliakov commented on the ongoing study of "Jewish blood" in an appendix entitled "The Origin of the Jews in the Light of Group Serology." Poliakov reiterated the fact that the four blood types were found among all peoples of the earth, but in different proportions that "correspond approximately to traditional racial classifications."[888] Aware of the anti-Semitic implications of such research, Poliakov explained, like so many before him, that the statistics for the blood type distributions of Jews throughout the Diaspora varied by country and were "remarkably close" to those of the people among whom they lived.[889] Overall, he explained, with the exception of several small Jewish "islets," the blood of Jews "corresponds to the European average." Like Ludwik Hirszfeld, Poliakov attributed this similarity to mixing in the first "thousand years of our era."[890] There was no unique "Jewish blood." At the same time, Poliakov acknowledged that blood-type surveys of the Jews were not "sufficiently numerous" and theorized that additional studies might well afford "enlightenment on certain details."[891] A decade later, Poliakov reported that there had been an "appreciable increase" in knowledge about serology; "although the findings do not clearly invalidate the earlier information," he explained, "they do suggest to scholars that considerable caution must be used in determining facts of Jewish history from biological data."[892] Others, too, refuted the significance of seroanthropology. In 1961 Professor A.E. Mourant, director of the Lister Institute of London, remarked at a conference on human population genetics held in Israel that "the study of the blood groups and other genetic characteristics of the Jews has thus far solved comparatively few problems."[893] The results were much the same with other racial types examined. Nonetheless, in the hopes of drawing some correlation, blood type surveys of Jews and other groups continued well into the 1980s.

---

[888] Leon Poliakov, *The History of Anti-Semitism*, Volume I: *From the Time of Christ to the Court Jews* (Philadelphia: University of Pennsylvania Press, 2003), 283.

[889] Ibid., 283–284.

[890] Ibid., 285.

[891] Ibid., 286.

[892] Ibid., 287.

[893] Ibid., 288.

As the continued interest in blood reveals, scientists suspected that blood was a bearer of genetic information long before this was proven. DNA was first discovered by Watson and Crick in 1953, a breakthrough that was reported in fewer than 1,000 words in the scientific journal *Nature*.[894] This knowledge was not applied until decades later, at which point DNA testing began to supersede the existing reliance upon the conventional four blood types. It was first used in the courts in the 1980s for paternity disputes and criminal cases. Instead of merely excluding a putative father, DNA could actually identify individuals. This trait could also be used in criminal cases to match perpetrators with crime scenes. Importantly, DNA has also changed anthropological analyses of blood, as it allows researchers to trace evolutionary history. By 1990 the International Human Genome Project had taken shape. Its main objective was to trace the genetic origins of various diseases in the hope of improving treatments, though it had definite racial implications. Through the collection of DNA samples, researchers hoped to decode the genetic material of the human race. Nations from across the globe joined, but Germany did not do so until 1995.[895] Blood and sperm samples were collected from a large number of donors. Rather than validate notions of "racial difference" in blood, the project seems to have "set us free from the obsessive notion that people have different and unequal racial characteristics in their blood."[896] As one author explains:

> Conclusive findings prove that our genome has no gene for race. Of the three billion letters in the human genome, 99.9 percent are the same in all human beings. The tiny remainder responsible for the differences between us often differs more between two people who are apparently similar and from the same culture group than between two people who look different and come from opposite ends of the world.[897]

In spite of this, controversy surrounding the genetic mapping of human beings continues, with the underlying fear that such knowledge might lead to some type of unjust discrimination. The 2008 work *Revisiting Race in*

---

[894] Joachim Pietzsch, "The World's Legacy Is in Our Blood: A Look at the First Century of Genetic Research," in James M. Bradburne, ed., *Blood: Art, Power, Politics, and Pathology* (New York: Prestel, 2002), 235.
[895] Ibid., 231.
[896] Ibid., 241.
[897] Ibid.

*a Genomic Age* draws attention to the fact that population geneticists argue that the genome holds the key to medically and forensically significant biological differences among human racial and ethnic populations. Increasingly, genetic variation among human populations—races, ethnicities, nationalities—is an object of keen biomedical interest.[898] These modern studies are not very different from their forerunners in the interwar period. Samples of blood are collected from large subject groups, and their genetic profile indicates their evolutionary origins—what some would consider to be merely a different phrase for "race." Furthermore, like the earlier seroanthropological blood type surveys, this research has definite political implications, and the terms "blood" and "race" continue to be based in shifting ideological presumptions.

---

[898] Barbara A. Koenig, Sandra Soo-Jin Lee, and Sarah S. Richardson, *Revisiting Race in a Genomic Age* (New Jersey: Rutgers University Press, 2008), 1.

# INDEX OF NAMES